SLOW

GROWS

THE

CHILD:

Psychosocial Aspects
of Growth Delay

Proceedings of a Symposium Sponsored by
The Human Growth Foundation and
Serono Symposia, USA

SLOW
GROWS
THE
CHILD:

Psychosocial Aspects of Growth Delay

Edited by
Brian Stabler
Louis E. Underwood
University of North Carolina

LEA LAWRENCE ERLBAUM ASSOCIATES, PUBLISHERS
1986 Hillsdale, New Jersey   London

Cover drawing by Brennan Stabler

Lawrence Erlbaum Associates, Inc., Publishers
365 Broadway
Hillsdale, New Jersey 07642

Library of Congress Cataloging in Publication Data
Main entry under title:
    Slow Grows the Child.

        Proceedings of a symposium held in Washington, D.C.
in Oct. 1984, and sponsored by the Human Growth
Foundation in collaboration with Serono Symposia USA.
        Includes index.
        1. Dwarfism—Psychological aspects—Congresses.
I. Stabler, Brian.   II. Underwood, Louis E.   III. Human
Growth Foundation.   IV. Serono Symposia USA.   [DNLM:
1. Growth Disorders—psychology—congresses.   WK 550
G8847 1984]
RJ135.G76   1986   616.4'7   85-13091
ISBN 0—89859-590-8

    Printed in the United States of America
    10 9 8 7 6 5 4 3 2 1

Contents

List of Contributors

Charles Annecillo, The Johns Hopkins University and Hospital, Baltimore, Maryland

Jennifer Bell, Columbia University College of Physicians and Surgeons

Robert M. Blizzard, University of Virginia School of Medicine, Charlottesville, Virginia

Richard R. Clopper, State University of New York at Buffalo and Children's Hospital of Buffalo

Carol Crouthamel, State University Hospital, Upstate Medical Center, Syracuse, New York

Heather J. Dean, University of Manitoba, Winnipeg, Manitoba

Sanford M. Dornbusch, Stanford University

Jennifer Downey, New York State Psychiatric Institute and Columbia University College of Physicians and Surgeons

Dennis Drotar, Case Western Reserve University School of Medicine

Paula M. Duncan, Stanford University

Anke A. Ehrhardt, New York State Psychiatric Institute and Columbia University College of Physicians and Surgeons

David G. Fish, University of Manitoba, Winnipeg, Manitoba

Henry G. Friesen, University of Manitoba, Winnipeg, Manitoba

Michael Gordon, State University Hospital, Upstate Medical Center, Syracuse, New York

Rhoda Gruen, New York State Psychiatric Institute and Columbia University College of Physicians and Surgeons

Clarissa S. Holmes, University of Iowa

Nancy J. Hopwood, C. S. Mott Children's Hospital, Ann Arbor, Michigan

Ann J. Johanson, University of Virginia School of Medicine, Charlottesville, Virginia

Susan Joyce, University of Virginia, Charlottesville, Virginia

Jennifer A. Karlsson, University of Iowa

Samuel Libber, Johns Hopkins University School of Medicine, Baltimore, Maryland

Margaret H. MacGillivray, State University of New York at Buffalo and Children's Hospital of Buffalo

Tom Mazur, State University of New York at Buffalo and Children's Hospital of Buffalo

Terri L. McTaggart, University of Manitoba, Winnipeg, Manitoba

Claude J. Migeon, Johns Hopkins University School of Medicine, Baltimore, Maryland

Barbara J. Mills, State University of New York at Buffalo and Children's Hospital of Buffalo

C. M. Mitchell, University of Virginia, Charlottesville, Virginia

Akira Morishima, Columbia University College of Physicians and Surgeons

Leslie Plotnick, Johns Hopkins University School of Medicine, Baltimore, Maryland

Ernest M. Post, State University Hospital, Upstate Medical Center, Syracuse, New York

Robert A. Richman, State University Hospital, Upstate Medical Center, Syracuse, New York

Philip L. Ritter, Stanford University

Ron G. Rosenfeld, Stanford University

Diane L. Rotnem, Yale University School of Medicine

Leonard P. Sawisch, Michigan Department of Education

Patricia Torrisi Siegel, Children's Hospital of Michigan, Detroit, Michigan

Jacqui Smith, Max-Planck-Institute for Human Development and Education, Berlin, West Germany

Hans-Christoph Steinhausen, Free University of Berlin, West Germany

Paul Tegtmeyer, State University Hospital, Upstate Medical Center, Syracuse, New York

Robert G. Thompson, University of Iowa

Mary L. Vorhess, State University of New York at Buffalo and Children's Hospital of Buffalo

Darrell M. Wilson, Stanford University

Deborah Young-Hyman, University of Maryland Medical School

Introduction:
Shortening the Odds in a Tall World

The odds that a short-statured person will be socially and emotionally fulfilled are judged by some to be not very good. There is a pervasive belief that equates tallness with strength, and shortness with weakness and a lack of social desirability. The recognition that delays in growth can be modified by medical therapies has led to increased awareness of psychological and social effects of short stature on children. To date, most reports on these effects have evolved from studies of the pathological elements of personality development in short individuals, through case studies or descriptive reports comparing short individuals with normals. In general, small groups of patients have been studied, retrospective reports have been used, and methods of study have been diverse, as have the backgrounds of researchers doing these investigations. As is often the case in clinical research, many investigators work in relative isolation, and publish their findings in disparate journals, chosen on the basis of their primary discipline. For these reasons, the study of psychological factors relevant to short stature and growth delay has progressed slowly. This slow progress was noted at a Human Growth Foundation Symposium held in Galveston, Texas, in 1979. The meeting, organized to bring together researchers and practitioners in the field, produced recommendations for future research and was the beginning of an informal network among researchers.

There has been little consensus about how best to measure the psychological and social adjustment of short individuals. One particularly troubling question is whether and how to use the instruments on growth-delayed patients that were standardized on psychiatric patients. Much of the earlier research on short stature tended to focus primarily on the maladaptation to shortness, relying on theoretical formulations of behavior that have their roots in mental illness, rather than in normal human development. The early reports would lead one to believe that short individuals are chronically depressed, dependent on others, socially isolated, and emotionally immature. These findings, which reflect the philosophical and theoretical orientations of the investigator, have proved not to be applicable to most short patients.

Another vexing aspect of this problem has been the focus on short stature as a single handicapping condition, and the resultant lumping of patients who have etiologies varying from constitutional short stature, to Turner syndrome, to chondrodysplasia. Not only are these conditions biologically distinct, but their psychological effects also are widely disparate. The term *short stature* describes a condition that in a given individual may change little over time. It implies certain social, environmental, and psychological sequelae. *Growth delay,* on the other hand, is the term used to describe a diminution of the normal rate of growth and maturation. It does not imply a fixed and unchanging condition. Because therapies may be available to modify growth rate, the patient and his family have hope for change, and the health professional is challenged by a variety of psychological and ethical issues. Whom should we treat? What is the aim of therapy? When should therapy be stopped? What do growth-delayed patients expect from treatment? All these are provocative questions yet to be answered by behavioral, social, and medical scientists. Clearly, the methods and instruments of clinical psychopathology will not provide many of the answers needed.

Because of these questions and others, The Human Growth Foundation, in collaboration with Serono Symposia USA, sponsored a symposium on psychosocial aspects of growth delay in Washington, DC, October 1984. Building on the experience of the meeting in Galveston over 5 years earlier, the meeting was structured so that researchers and practitioners in medicine, psychology, and social work could present their work. Experts from a variety of disciplines had the opportunity to discuss their experience in groups of limited size. A number of individuals bore special responsibility for organizing the symposium; among them are Denise Orenstein, Executive Director of the Human Growth Foundation, and Leslie Nies and her very capable staff at Serono Symposia USA. They deserve special thanks for their tireless efforts in planning the symposium and making it run so smoothly. Ann Johanson, M.D., a member of the Human Growth Foundation Board of Directors, was instrumental in bringing many of the contributing participants to our attention, and in selecting appropriate resources. The individuals who served as core faculty shared the material that made the symposium such a success, and their contributions can be assessed in the pages that follow. Our hope is that these Proceedings will advance understanding of the social and psychological experience of growth delay, and increase the odds that our medical and psychological intervention will produce the most desirable outcome.

Brian Stabler
Louis E. Underwood

1 Longitudinal Evaluation of Behavior Patterns in Children with Short Stature

Clarissa S. Holmes
Jennifer A. Karlsson
Robert G. Thompson
University of Iowa

The potential availability of synthetic human growth hormone, obtained using recombinant DNA technology, has fostered increased interest in psychological sequelae that may be associated with short stature. Short children with no demonstrable endocrine disorder have been excluded from the very limited supplies of growth hormone harvested from cadavers, but they are being considered as viable recipients of the more abundant synthetic product (Gertner et al., 1984, Underwood, 1984). Preliminary medical data show synthetic growth hormone to be effective in promoting growth in these children over brief periods of time (Gertner et al., 1984); however, the long-term physiological risks and benefits remain unknown (Underwood, 1984). Psychological benefits of enhanced growth will likely be a factor in the decision to begin medical treatment, perhaps analogous to intervention considerations associated with alteration of body appearance through plastic surgery.

Current information about psychological adjustment to short stature is derived primarily from research with pediatric populations who have experienced growth retardation at least two standard deviations below age expectation. Short children have been characterized as experiencing more internalizing behavior problems than peers of average stature. Specifically, greater somatic complaints and withdrawal (Gordon, Crouthamel, Post, & Richman, 1982; Holmes, Hayford, & Thompson, 1982b) and less aggressive and dominant behavior (Steinhausen & Stahnke, 1976) have been reported. Children with growth hormone deficiency (GHD) may be less able to cope adaptively with frustration than taller age-mates because of their tendency to internalize negative feelings (Drotar, Owens, & Gotthold, 1980). Not all

1

internalizing behaviors differentiate GHD children from controls, as Stabler and Underwood (1977) report similar response patterns between GHD children and normals on measures of locus of control and anxiety. Behavior patterns undoubtedly vary somewhat as a function of situation. Parent ratings of behavior at home have indicated high levels of both internalizing and externalizing behavior problems, whereas teachers report more internalization at school (Holmes, Hayford, & Thompson, 1982a).

Academic and intellectual functioning also has been a focus of evaluation in populations of short children. Results indicate that overall intellectual development is not retarded in conjunction with delayed physical maturation. Despite average cognitive abilities, several studies have reported significant school problems, thought secondary to emotional distress. Noted are difficulties with academic underachievement and high incidences of grade retention (23% to 41%) in children with GHD (Holmes, Thompson, & Hayford, 1984; Pollitt & Money, 1964; Steinhausen & Stahnke, 1976) and girls with Turner's Syndrome (TS; Holmes et al., 1984); the latter also experience well-documented visual spatial and math deficits (e.g., Hier, Atkins, & Perlo, 1980). In contrast, the academic standing of children with constitutional delay (CD) appears more variable. Gordon, Post, Crouthamel, and Richman (1984) report no academic delays, whereas Gold (1978) and Steinhausen and Stahnke (1976) both report underachievement and, in the latter study, a very high incidence of grade retention (60%) is reported. Age-related effects were obtained by Holmes et al. (1982a) who found that older CD children experience school difficulties but younger CD children do not.

Perhaps patterns of school and behavioral functioning in short children will become clearer with further evaluation of relevant subject characteristics such as gender and age. To date, there has been a tendency to collapse data over large age spans and across sex; to assess restricted age ranges; or to fail to indicate the age and sex composition of study samples. Patterns of functioning also may be better elucidated by reducing sampling variability with periodic longitudinal assessment of the same subjects. This was the purpose of the present longitudinal evaluation.

METHOD

Subjects

Forty-seven of 76 children sought for follow-up participated in this longitudinal study, approximately 3 years after their initial evaluation. Partial data from 23 subjects (initial evaluation only) have been reported previously by the investigators (Holmes et al., 1982a, 1982b). All children were outpatients at a university pediatric endocrinology clinic, having been referred by their local physicians. When first enrolled in the study, all children were a minimum of two standard deviations below height expectation for age and sex, according to the National Center for Health Statistics (1977). Etiology of short stature was due to GHD ($N = 17$), CD ($N = 21$), or TS ($N = 9$). The medical evaluation of short stature included a detailed family history to document family growth patterns, gestational and perinatal history to assess intrauterine growth and development, and a complete physical examination with emphasis on the assessment of a series of anthropometric indices of long bone and axial skeletal growth. Appropriate laboratory evaluations to exclude renal or thyroid dysfunction were routinely completed. Bone maturation was assessed by the method of Greulich and Pyle (1959). A diagnosis of constitutional delay of growth was made on the basis of documented growth velocity within two standard deviations of normal velocity for age, significant retardation of bone age that was compatible with other physical parameters of body maturation, and projected adult height estimated by the method of Bayley and Pinneau (1959) within normal limits. Although a family history compatible with constitutional delay was frequently obtained, its presence was not a criterion for diagnosis. Subjects with historical or physical findings compatible with hypopituitarism or with significantly reduced growth velocity had assessment of the integrity of growth hormone and adrenocorticotrophic hormone secretion with an arginine-insulin tolerance test (Penny, Blizzard, & Davis, 1969). Growth hormone deficiency was diagnosed if maximum growth hormone concentration failed to exceed 7 ng/ml on any of the nine sequential samples obtained.

Previous intellectual screening with the Vocabulary and Similarities subtests from the Wechsler Intelligence Scale for Children-Revised (WISC-R) (Wechsler, 1974) indicated all children possessed at least average intelligence. Socioeconomic information based on the Hollingshead's (1961) Four Factor Index was also obtained at time of first evaluation and indicated a predominantly middle-class socioeconomic background. These measures were not repeated at follow-up because of the relative stability of social class and IQ scores over time. Based on age at initial evaluation, the sample of children was divided into older and younger age groups for longitudinal follow-up. See Table 1.1.

Assessment Measures

At the time of initial and follow-up evaluations, parents were asked to complete an Achenbach (1979) Child Behavior Problem Checklist (CBPC),

TABLE 1.1
Mean Scores of Short Children on Demographic and
Medical Variables When Grouped by Age

	Age	
	Younger (N = 22)	Older (N = 25)
Age at Initial Evaluation	9.7	14.4
	(1.9)	(1.6)
Age at Follow-up Evaluation	12.8	17.5
	(2.1)	(2.0)
Prorated Verbal IQ	101.6	101.7
	(13.4)	(13.8)
Height Age (HA) at Follow-up	8.1	11.8
	(2.3)	(1.5)
CA/HA Discrepancy	3.5	4.2
	(1.3)	(1.5)
Tanner Stage at Follow-up	1.0	2.4
	(1.3)	(1.4)

Note.   Figures in parentheses are SDs.

describing their child's behavior. The first portion of the CBPC assesses skills that are divided into three scales: Activities, Social, and School Competence. The Activities (ACT) scale measures the frequency and skill with which a child engages in various sports, hobbies, and household chores. The Social (SOC) scale measures the frequency and proficiency of a child's social involvements (e.g., participation in clubs, socializing with friends). An estimate of the child's academic performance (e.g., grade point average, need for remedial education, or retention) is provided by the School (SCH) scale. Performance on all three Competence scales is tabulated as normalized T scores ($\overline{X}$ = 50, SD = 10), with higher scores indicating greater competence.

The second portion of the CBPC assesses Behavior Problems. A variety of common childhood behavior problems have been factor analyzed, yielding two broad subtypes, Internalizing and Externalizing tendencies. The former refers to inhibited, anxious behaviors (e.g., somatic complaints, withdrawal, and obsessiveness), whereas the latter refers to more acting out or aggressive behavior problems (e.g., delinquency, cruelty, or hyperactivity). The Behavior Problem scales yield normalized T scores ($\overline{X}$ = 50, SD = 10). However, unlike the Competence scales, higher scores on the Internalizing (INT) and Externalizing (EXT) scales indicate greater behavioral disturbance.

In addition to the CBPC, medical information was obtained from each child's most recent pediatric endocrinology evaluation. Specifically, height information (percentile rank, height age) and an estimate of pubertal development (i.e., Tanner Stage) were obtained from medical records.

RESULTS

At follow-up, 3.1 years after initial evaluation (range = 2.2 to 4.0 years), the majority of short children remained at less than the 5th percentile for height. Tables 1.1 and 1.2 provide descriptive data about short children when grouped according to age and etiology of short stature, respectively. Separate one-way analysis of variance failed to reveal significant differences in prorated IQ scores, regardless of subject classification. When grouped by etiology of short stature, height age was similar for all groups, but the discrepancy between chronological age and height age (CA/HA) indicated GHD children experienced the greatest growth retardation and CD children the least. Sexual development, indicated in Tanner stages, was equivalent for GHD, CD, and TS groups. Of the 16 children identified as growth hormone deficient, 8 had multiple pituitary trophic hormone deficiencies.

A preliminary $2 \times 2 \times 3 \times 2$ MANOVA with one repeated factor (Trial) and three between subject factors (Sex × Diagnosis × Age) was conducted on scores from the five scales of the CBPC (ACT, SOC, SCH, INT, EXT) to examine sex-related effects. A significant interaction of Diagnosis × Sex was obtained, Wilk's $F(5.26) = 2.48$, $p < .06$. The univariate analyses

TABLE 1.2
Mean Scores of Children on IQ and Medical Variables
When Grouped by Etiology of Short Stature

	Etiology of Short Stature		
	GHD[a] (N = 17)	CD[b] (N = 21)	TS[c] (N = 9)
Prorated Verbal IQ	97.3	104.1	102.9
	(12.5)	(14.2)	(13.6)
Height Age (HA) at Follow-up	9.8	10.1	10.6
	(3.2)	(2.8)	(1.5)
CA/HA Discrepancy	4.7	3.1	3.9
	(1.5)	(.9)	(1.3)
Tanner Stage	1.5	1.9	1.9
	(1.6)	(1.4)	(1.6)

Note.   Figures in parentheses are SDs.
[a]GHD = Growth Hormone Deficiency
[b]CD = Constitutional Delay
[c]TS = Turner's Syndrome

revealed a significant effect for the SCH variable, $F(1,30) = 4.21, p < .05$. Tukey follow-up tests were employed for this and all other multiple comparisons, with significant results reported at the $p < .05$ level. Results showed CD females obtained significantly better scores than GHD and TS girls, whose scores did not differ (See Fig. 1.1). Based on previous findings of an Age × Sex interaction (Holmes et al., 1982a) an a priori decision was made to evaluate this effect, although the MANOVA interaction was not significant. At the univariate level, a significant effect, $F(1,30) = 5.27, p < .03$ was obtained for the SCH variable. Figure 1.2 shows results of the multiple comparisons.

As the factor of Sex failed to yield further significant MANOVA results, the data from males and females were combined for subsequent analyses. To control Type I error, a 2 × 3 × 2 MANOVA with one repeated factor (Trial) and two between subject factors (Diagnosis × Age) was then conducted with the five dependent variables. Significant Age × Trial, Wilk's $F(3,35) = 3.73, p < .009$, and Diagnosis × Age, Wilk's $F(10,74) = 1.74, p < .09$, interactions were found along with a significant main effect for Age, Wilk's $F(5,37) = 3.37, p < .02$. In univariate analyses, the Age × Trial interaction was significant for both the SOC, $F(1,39) = 10.76, p < .003$, and the SCH $F(1,39) = 4.52, p < .04$, variables. Follow-up tests on SCH scores failed to

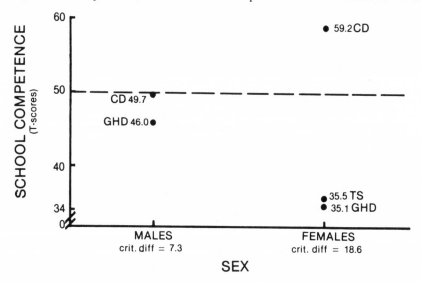

FIG. 1.1.   Normalized T scores ($\overline{X} = 50, SD = 10$) indicating mean level of school competence. Critical differences (crit. diff.) must be exceeded for multiple comparisons to be significant at the $p < .05$ level. Mean scores are reported for males and females with growth hormone deficiency (GHD; crit. diff. = 8.4) and constitutional delay (CD; crit. diff. = 12.1); the absence of males with Turner's syndrome (TS) obviates a sex comparison.

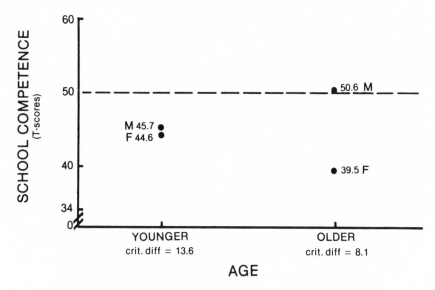

FIG. 1.2.   Mean *T* scores indicating level of school competence for older and younger females ("F"; crit. diff. = 14.0) and older and younger males ("M"; crit. diff. = 7.1).

yield significant differences; Fig. 1.3 graphically reports these data. Further evaluation of the SOC scores revealed significantly different multiple comparisons as illustrated in Fig. 1.4. At the univariate level, the Diagnoses × Age interaction was significant for the SCH variable, $F(2,41) = 3.32, p < .05$. Follow-up comparisons are illustrated in Fig. 1.5. Univariate analyses also yielded a significant main effect of Age for both the INT, $F(1,41) = 7.51, p < .009$, and EXT, $F(1,4) = 10.22, p < .003$, behavior scales; see Fig. 1.6 for an illustration of follow-up results.

DISCUSSION

Longitudinal data from the present study suggest that children with short stature appear to undergo an age-related decline in adjustment during early adolescence.Specifically, as Figs. 1.3 and 1.4 show, parental ratings of school and social competence fall for both younger and older groups of children at 12 and 14 years, respectively. The typical physical, social, and school-related changes that occur during preadolescence may be particularly stressful when children have a chronic medical condition, such as short stature, which visibly differentiates them from age-mates. These years of relative difficulty, with performance about one standard deviation below the mean, are preceded and followed by functioning near the 50th percentile.

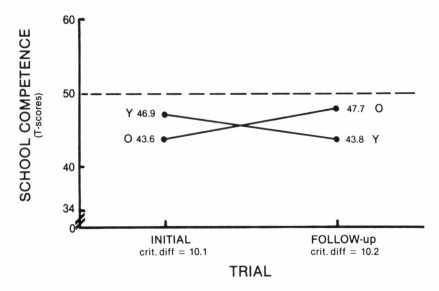

Fig. 1.3.   Mean *T* scores indicating level of school competence at initial and follow-up evaluations for younger ("Y"; crit. diff. = 4.7) and older ("O"; crit. diff. = 4.5) subjects.

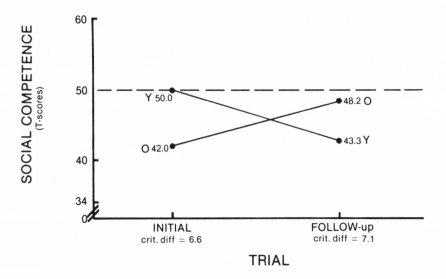

FIG. 1.4.   Mean *T* scores indicating level of social competence at initial and follow-up evaluations for younger ("Y"; crit. diff. = 4.7) and older ("O"; crit. diff. = 5.7) children.

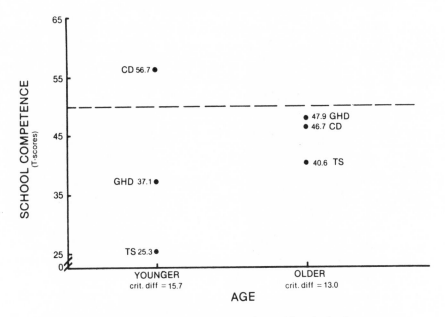

FIG. 1.5.   Mean *T* scores indicating level of school functioning for younger and older GHD (crit. diff. = 7.6), CD (crit. diff. = 10.1), and TS (crit. diff. = 19.9).

Fortunately, it appears that children rebound from the turmoil of early adolescence rather than becoming enmeshed in a cycle of continuing adjustment problems. Competencies in school and social functioning return to age expectation by age 17. This finding, coupled with age-expected competency ratings at age 9, suggest that the disequilibrium associated with early adolescence is indeed a temporary transition. However, this conclusion is derived from both a longitudinal and cross-sectional inspection of the data; further follow-up of the younger group is needed to yield more definitive information about the course of adjustment in the same sample of short children.

School performance, as assessed by the CBPC, was also associated with several sex-related effects. First, as Fig. 1.2 illustrates, older females in this sample had more school problems than other groups of short children. This finding is consistent with an earlier report (Holmes et al., 1982a). Closer evaluation of specific items endorsed by parents on the School scale indicates that many adolescent females suffer a constellation of problems including poor grades, enrollment in remedial or special classes, and a relatively high (approximately 25%) likelihood of grade retention during

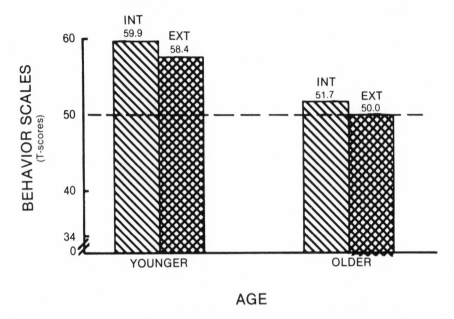

FIG. 1.6.   Mean T scores for younger and older subjects on the internalizing (INT) behavior scale (crit. diff. = 5.7) and externalizing (EXT) behavior scale (crit. diff. = 5.0).

their academic career. As Fig. 1.1 suggests, school problems experienced by females are largely accounted for by girls with GHD and TS, who performed 1½ standard deviations below the mean. As Table 1.2 indicates, children with GHD have the greatest discrepancy between chronological age and height age compared to other groups, being almost 5 years delayed in expected height age. Females rather than males appear more sensitive to physical deviations in height, perhaps related to greater social concerns. Girls with TS in the present sample experienced extremely high incidences of poor grades (71%) for which they were likely to be retained at least one grade (63%) and/or to receive special class placement (59%). These problems are probably secondary to well-documented visual spatial and math deficits (e.g., Hier et al., 1980).

In contrast to the academic problems of adolescent females, parents indicated that boys, and children with CD, were functioning academically near age expectancy (Figs. 1.1 and 1.2). However, an age effect revealed younger CD children to have significantly better academic performance, over one standard deviation (SD = 10) higher, than all other groups of children (Fig. 1.5). This pattern of results is consonant with findings from an earlier study (Holmes et al., 1982a). An age-related trend may explain why Gordon et al. (1984) found age-appropriate academic functioning in their sample composed entirely of younger children ($\overline{X}$ age = 10.4), whereas Steinhausen and Stahnke (1977) studying an older sample ($\overline{X}$ age = 14.9 years) found academic underachievement. These age-specific differences in academic skills underscore the need to systematically evaluate age as a subject variable to reconcile and better understand performance discrepancies reported across studies.

Younger short children displayed more behavior problems than their older counterparts. As Fig. 1.6 shows, significantly more internalizing problems, characterized by withdrawal and somatic complaints, were noted. Similarly, younger short children experienced significantly higher levels of externalizing behaviors, characterized by aggressiveness, delinquency, and/or hyperactivity, whereas the behavioral profiles of older children showed a balance of internal and external behaviors, clearly within age expectation. The more subdued behavior of older children is undoubtedly related to improvements found in social and academic competency at time of follow-up evaluation (see Figs. 1.3 and 1.4).

The present data indicate the utility of a longitudinal paradigm to better understand the course of children's adjustment to short stature. Even though subject attrition may produce a follow-up sample comprised of less favorable responders to medical treatment, one could argue that these children are the most suitable subjects in whom to examine psychological sequelae associated with chronic shortness. Despite this possible sample bias in the present study, results indicated that older children nevertheless became more adept at accommodating to their environment. The present results also highlight the need for future research to systematically study subject variables such as age, sex, and etiology of short stature.

REFERENCES

Achenbach, T.M. (1979). The child behavior profile: An empirically based system for assessing children's problems and competencies. *International Journal of Mental Health, 7,* 24–42.

Bayley, N., & Pinneau, S. R. (1959). Tables for predicting adult height from skeletal age: Revised for use with the Greulich-Pyle hand standards. In W. W. Greulich & S. L. Pyle (Eds.), *A radiographic atlas of skeletal development of the hand and wrist* (2nd ed., pp. 231–250). Stanford: Stanford University Press.

Drotar, D., Owens, R., & Gotthold, J. (1980). Personality adjustment of children and adolescents with hypopituitarism. *Child Psychiatry and Human Development, 11,* 59–66.

Gertner, J. M., Genel, M., Gianfredi, S. P., Hintz, R. L., Rosenfeld, R. G., Tamborlane, W. V. & Wilson, D. M. (1984). Prospective clinical trial of human growth hormone in short children without growth hormone deficiency. *Journal of Pediatrics, 104,* 172–176.

Gold, R. F. (1978). Constitutional growth delay and learning problems. *Journal of Learning Disabilities, 11,* 36–38.

Gordon, M., Crouthamel, C., Post, E. M., & Richman, R. A. (1982). Psychosocial aspects of constitutional short stature: Social competence, behavior problems, self-esteem, and family functioning. *Journal of Pediatrics, 101,* 477–480.

Gordon, M., Post, E. M., Crouthamel, C., & Richman, R. A. (1984). Do children with constitutional delay really have more learning problems? *Journal of Learning Disabilities, 17,* 291–293.

Greulich, W. W., & Pyle, S. I. (1959). *A radiographic atlas of skeletal development of the hand and wrist* (2nd ed.). Stanford: Stanford University Press.

Hier, D. B., Atkins, L., & Perlo, V. P. (1980). Learning disorders and sex chromosome aberrations. *Journal of Mental Deficiency Research, 24,* 17–26.

Hollingshead, A. B., & Redlich, F. C. (1958). *Social class and mental illness: A community study.* New York: Wiley.

Holmes, C. S., Hayford, J. T., & Thompson, R. G. (1982a). Parents' and teachers' differing views of short children's behavior. *Child Care, Health and Development, 8,* 327–336.

Holmes, C. S., Hayford, J. T., & Thompson, R. G. (1982b). Personality and behavior differences in groups of boys with short stature. *Children's Health Care, 11,* 61–64.

Holmes, C. S., Thompson, R. G., & Hayford, J. T. (1984). Factors related to grade retention in children with short stature. *Child: Care, Health and Development, 10,* 199–210.

National Center for Health Statistics. (1977). *Growth curves for children: Birth–18 years.* Hyattsville, MD: US Department of Health, Education and Welfare.

Penny, R., Blizzard, R. M., & Davis, W. T. (1969). Sequential arginine and insulin tolerance test on the same day. *Journal of Clinical Endocrinology and Metabolism, 29,* 1499–1501.

Pollitt, E., & Money, J. (1964). Studies in the psychology of dwarfism. I. Intelligence quotient and school achievement. *Journal of Pediatrics, 64,* 415–421.

Stabler, B., & Underwood, L. E. (1977). Anxiety and locus of control in hypopituitary dwarf children. *Research Relating to Children Bulletin, 38,* 75.

Steinhausen, H., & Stahnke, N. (1976). Psychoendocrinological studies in dwarfed children and adolescents. *Archives of Disease in Childhood, 51,* 778–783.

Steinhausen, H., & Stahnke, N. (1977). Negative impact of growth hormone deficiency on psychological functioning in dwarfed children and adolescents. *European Journal of Paediatrics, 126,* 263–270.

Underwood, L. E. (1984). Growth hormone treatment for short children. *Journal of Pediatrics, 104,* 237–239.

Wechsler, D. (1974). *Wechsler Intelligence Scale for Children—Revised.* New York: Psychological Corporation.

2 Academic and Emotional Difficulties Associated with Constitutional Short Stature

Robert A. Richman

Michael Gordon

Paul Tegtmeyer

Carol Crouthamel

Ernest M. Post
State University Hospital, Upstate Medical Center, Syracuse, New York

It has long been assumed that children with short stature have more social, academic, and psychological problems than those with normal stature (Drash, 1969; Gold, 1978; Money & Pollitt, 1966; Pollitt & Money, 1964). They were believed to experience low self-esteem, a high degree of social isolation, withdrawal, immaturity, and disturbances of body image. However, recent studies have shown that growth hormone deficient children had no greater occurrence of abnormalities in general psychological adjustment, sex role development, body image (Drotar, Owens, & Gotthold, 1980), anxiety, or locus of control (Stabler & Underwood, 1977). Although such children seem to have normal psychological adjustment, they represent less than 1% of the children with growth disorders (McArthur & Fagan, 1971). Growth hormone deficient children probably experience short stature differently from other short children, because their condition is being ameliorated by medical therapy. Most short children have constitutional short stature (Horner, Thorsson, & Hintz, 1978), which is characterized by (a) normal birth weight; (b) growth failure between 6 months and 3 years; (c) subsequent normal growth velocity; but (d) height below the 5th percentile for age throughout childhood. This condition is also referred to as constitutional growth delay and constitutional delay of growth and adolescence. Because the period of growth failure is associated with a deceleration in skeletal maturation, the potential for attaining normal adult height remains excellent. The onset of pubertal changes, including the growth spurt, is usually delayed. Children with constitutional short stature usually become adults of normal stature, but may suffer lasting psychologic effects from their previous short stature.

To expand our knowledge concerning the academic and emotional concomitants of constitutional short stature, we have performed a series of studies of children with constitutional short stature (Gordon, Crouthamel, Post, & Richman, 1982; Gordon, Post, Crouthamel, & Richman, 1984). We administered a battery of questionnaires and psychological and academic tests to assess personality and intellectual, visual-motor, and academic performance. Our objective was to determine whether children with constitutional short stature develop personality styles that adversely affect emotional development, social behavior, and academic performance.

METHODS

We studied 24 children, 20 boys and 4 girls, with constitutional short stature who had been evaluated in the Pediatric Endocrinology Center, SUNY Upstate Medical Center, Syracuse, NY Their ages ranged from 6 to 13 years (10.4 ± 0.4 years; $\bar{x}$ ± SEM [standard error of the mean]). Their heights were below the 5th percentile for age and sex (National Center for Health Statistics, 1977); skeletal maturation (Greulich & Pyle, 1959) was at least 13 months less than chronological age; height velocity was greater than 4 cm per year; and laboratory test results were normal. For comparison, we also tested 23 healthy children with normal heights and no history of psychiatric disturbances. They were selected from the roster of a private pediatric practice, and were matched to our subjects for age (10.5 ± 0.4 years), Full Scale IQ (109.8 ± 2.1; Wechsler Intelligence Scale for Children-Revised (WISC-R; Wechsler, 1974), socioeconomic status (2.9 ± 0.2; Hollingshead & Redlich, 1958), and sex (Gordon et al., 1984). None of the children in either group had a Full Scale IQ in the retarded or superior ranges.

One of two certified school psychologists administered eight subtests of the WISC-R (Wechsler, 1974), the Peabody Individual Achievement Test (Dunn & Markwardt, 1970), and the Bender–Gestalt Test (Bender, 1938) which was scored according to Koppitz (1964). These tests assessed intellectual functioning, academic achievement, and visual-motor integration, respectively. Parents were asked to complete the Child Behavior Checklist (Achenbach, 1979), from which we generated indices of overall social competence and behavior problems. This checklist estimates the degree to which a child internalizes difficulties, by quantifying somatic complaints, social withdrawal, and depression. It also measures externalizing behavior such as hyperactivity, aggression, and fabrication. Parents also completed the Maryland Parent Attitude Survey (Pumroy, 1966) and the Family Functioning Index (Pless & Satterwhite, 1973) to provide information about child-rearing attitudes and family functioning. Finally, parents completed our questionnaire regarding developmental, academic, and psychiatric history. Among the questions were (1) Has your child ever had academic or learning problems? (2) Has your child ever repeated a grade or been placed in a special class? After obtaining parental consent, teachers were requested to complete the Teacher Form of the Child Behavior Checklist (Achenbach & Edelbrock, 1980), and the 39-item Teacher Rating Scale (Conners, 1969). The children completed the Piers-Harris Self-Concept Scale (Piers, 1969), which provides information concerning their view of their behavior, intellectual status, popularity, physical appearance, and levels of anxiety and happiness.

To determine the degree to which an individual perceives reinforcements for behavior as noncontingent and attributable to unpredictable forces, such as chance, fate, or actions by powerful others, we administered the Nowicki–Strickland Locus of Control Scale for Children (Nowicki & Strickland, 1973). This scale measures feelings of helplessness and hopelessness, or, conversely, of being in control of one's fate.

To assess general personality development, we administered the Rorschach Inkblots, scored according to the Comprehensive System (Exner, 1974). For this particular study, we departed from standard procedure by limiting each subject to two responses per card. In addition to the standard structural determinants, protocols were also scored for aggression (Crain & Smoke, 1981), hostility (Goldfried, Stricker, & Weiner, 1971, pp. 89–116), oral-dependency (Masling, Rabie, & Blondheim, 1967), and body boundary (Fisher & Cleveland, 1968).

Body image was assessed by analyzing human figure drawings in the Draw-a-Person Test using standard procedures (Machover, 1949). Subjects were requested to draw on an $8^{1}/2 \times 11$ inch piece of plain white paper any person they wished, a person of the opposite sex to that of their first drawing, and a self-portrait. The drawings were scored for developmental maturity according to the Koppitz criteria (1968). This method considers features such as integration, slanting figures, transparencies, asymmetry, shading, and omission of body parts. In addition, a sum omission score was given based upon the total of all of the omissions of the aforementioned features. The scores for each of these features were the sums across the three drawings. The first two figures (a male–female pair) were also scored for the level of sexual differentiation (Haworth & Normington, 1961). We also recorded the sex of the first-drawn figure, sex of the taller figure, and sex of the more elaborate figure (according to the criteria of Haworth & Normington, 1961). All protocols were scored by an experienced clinician who was unaware of the subject's identity. A second rater then scored the drawings. The agreement between raters was high for all scores generated (mean interrater reliability r = .92). Sex role development was measured using the Haworth–Normington (1961) method.

The Thematic Apperception Test (Murray, 1943) was administered to assess thought content and areas of psychologic conflict. Children were asked to tell stories based upon eight TAT pictures (Cards 1, 2, 3BM, 8BM, 13B, 18GF, 11, 17BM, and 19), and their responses were scored for the presence or absence of imagery related to the need for power (Veroff, 1957), achievement (McClelland, Clark, Roby, & Atkinson, 1949), and affiliation (Shipley & Veroff, 1952).

RESULTS

The scores of the children with constitutional short stature on tests of intellectual functioning, academic achievement, and visual-motor performance were not statistically different from those of the children with normal stature for any of the 15 variables tested (Gordon et al., 1984). Likewise, the teachers' ratings of school adjustment, as measured by the Teacher Form of the Child Behavior Checklist and the Teacher Rating Scale, showed no differences between the two groups (Gordon et al., 1984). According to the parents, about one third of both groups had academic difficulties. However, five of the children with constitutional short stature had repeated a grade, whereas only one of the children in the control group had (p = .15, Fisher Exact Test).

The parents of the children with short stature evaluated their children quite differently than did the teachers (Gordon et al., 1982; Gordon et al., 1984). Based upon the parents' ratings, both groups of children had comparable scores on the social competence factor, $t(40)$ = 1.5, p < .1, suggesting similar school performance and involvement in activities (Fig. 2.1). The children with constitutional short stature did participate in fewer social groups. They also differed significantly on the Behavior Problem Index, $t(40)$ = 1.9, p < .05, having more somatic complaints, $t(42)$ = 1.9, p < .05, schizoidal tendencies, $t(41)$ = 2.7, p < .025, and social withdrawal, $t(27)$ = 2.3, p < .025 (Figs. 2.2 & 2.3). The children with short stature had elevated scores similar to those typically found in children referred for psychiatric evaluation (Hollingshead & Redlich, 1958, pp. 398–407). There were also significant differences between the parent ratings of children with short stature and those of normal children on the Maryland Parent Attitude Survey and the Family Life Survey (Fig. 2.4; Gordon et al., 1982). Parents of the children with short stature tended to set less clear limits on behavior, $t(39)$ = 2.2, p < .05. These families also scored significantly lower in

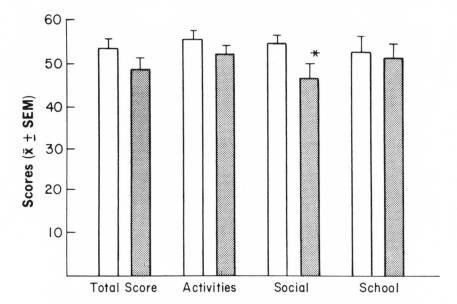

FIG. 2.1. Parents' ratings of social competence on the Child Behavior Checklist. Normal stature, open bars; constitutional short stature, hatched bars. * indicates $p < .05$.

cooperation and effective communication ($t(37) = 2.1, p < .05$) on the Family Life Survey.

Unlike the teachers whose ratings did not distinguish between short and normal children, the subjects rated themselves as being different. Children with constitutional short stature rated themselves as having lower self-esteem than their peers with normal stature. They also saw themselves as unhappy, $t(45) = 2.1, p < .05$, and unpopular, $t(45) = 2.1, p < .05$; (Fig. 2.5). These responses were not influenced by age or sex. However, there were no differences between the two groups as to the locus of control. Both groups perceived reinforcements for their behavior as contingent upon and attributable to predictable forces. Neither group exhibited increased feelings of helplessness and hopelessness.

Although there were only a few (6 of 58 scores) significant differences in general psychologic development as measured by the Rorschach Test, there was a logical pattern to them. Protocols of children with constitutional short stature were suggestive of a more introverted personality style. They contained more human movement than normal children. They had more morbid, $t(45) = 2.2, p < .03$, oral dependent, $t(45) = 2.3, p < .03$, and barrier content, $t(45) = 2.3, p < .03$, scores. They also had higher percentage of "good form responses."

The perceptions of their bodies by children with constitutional short stature as measured by the Draw-A-Person test also differed from children

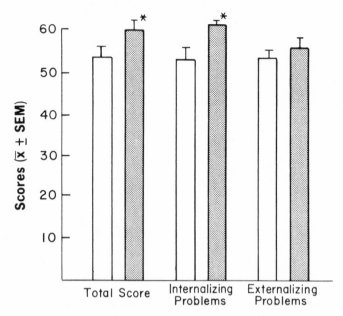

FIG. 2.2.   Parents' ratings of behavior problems on the Child Behavior Checklist. Normal stature, open bars; constitutional short stature, hatched bars. * indicates $p < .05$.

with normal stature. The mean height (10.88 ± 0.96 cm) of the self-drawing for children with constitutional short stature was not significantly different from (10.03 ± 0.96 cm) the normal children. The size of the figures were not influenced by the gender depicted. The total number of abnormal structural features per figure was too small to make valid inferences. The total number of abnormalities with respect to sexual differentiation, for all three figures drawn by each child, did not differ between the two groups. Short children did not tend to draw the opposite-sex figure first, larger, or in a more elaborate fashion than the same-sex figure. However, differences were found for two of the six major structural features scored. Children with constitutional short stature preferred to use shading, $t(42) = 3.07$, $p < .01$, and to omit body parts, $t(41) = 2.32$, $p < .05$. The frequency of omission of any one particular body part was very low, often occurring only once for normal children. Short children omitted each of the eight individual body parts scores more frequently than the normal children. The differences were significant for the omission of hands, $X^2 = 3.72$, $p < .05$, neck, $X^2 = 4.97$, $p < .05$, and feet, $X^2 = 8.73$, $p < .01$. Sum omissions correlated inversely with the age of the children for the combined groups, $r = -.55$, $p < .01$.

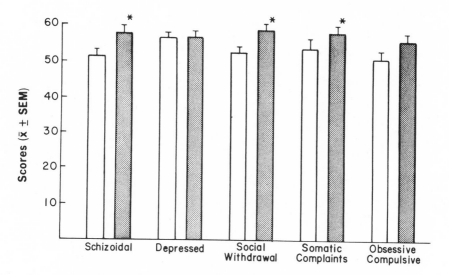

FIG. 2.3.   Parents' ratings of internalizing problems on the Child Behavior Checklist. Normal stature, open bars; constitutional short stature, hatched bars. * indicates $p < .05$.

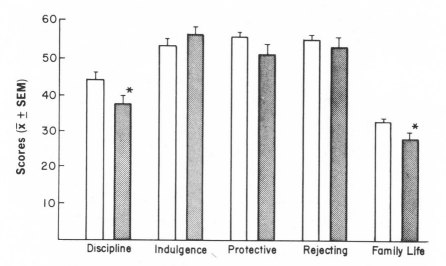

FIG. 2.4.   Parents' attitudes about child rearing and family life. Parents were given the Maryland Parent Attitude Survey and Family Life Survey. Normal stature, open bars; constitutional short stature, hatched bars. * indicates $p < .05$.

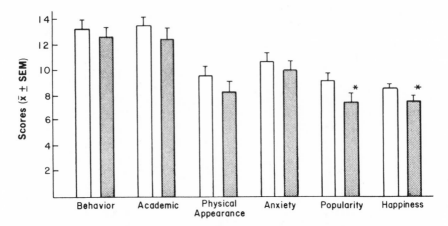

FIG. 2.5.   Children's ratings on the Piers-Harris Self-Concept Scale. Normal
stature, open bars; constitutional short stature, hatched bars. * indicates $p <$
.05.

We attempted to validate the interpretation of the omission of body part
scores by correlating them individually with the behavior problem factors
on the Child Behavior Checklist (Table 2.1). Although there were only 10
significant associations between the variables, certain patterns did appear.
Those subjects who omitted eyes tended to have schizoid tendencies, $r =$
.35, $p < .02$, and exhibited signs of depression. Omission of the nose
occurred in uncommunicative, $r = .33, p < .03$, individuals with obsessive-
compulsive traits, $r = .44, p < .003$. Children omitting both arms and legs
had scored significantly lower for depression and aggression, whereas those
who omitted both feet or the torso were likely to be more hyperactive, $r =$
.32, $p < .04$, and $r = .37, p < .03$, respectively. Hyperactivity was also
increased in those children with high omissions scores, $r = .39, p < .01$.

DISCUSSION

Our results form a relatively coherent pattern of the adjustment and
personality characteristics that typify our group of subjects with constitutional short stature. We have found, using standardized test instruments,
that such children do not differ from their peers with normal stature in
intellectual function, academic achievement, and visual-motor integration.
They also had comparable scores to the normal children on the Social
Competence factor, suggesting no difference in school performance or
involvement in activities. Because teacher behavior ratings were also quite
similar for the two groups, growth failure did not seem to appreciably affect

TABLE 2.1

Correlations Between Omissions of Critical Body Parts and
Parents' Ratings on Child Behavior Checklist Factors (N = 47)

	Schizoid Tendencies	Depression	Uncommunicative Behavior	Obsessive Compulsive Behavior	Aggression	Hyperactivity
Eyes	.35*	.35*	.27	.22	−.05	.22
Nose	.08	.11	.33*	.44*	.11	.18
Arms	−.16	−.31*	−.20	−.13	−.31*	.21
Hands	−.12	.10	.05	.24	.01	.28
Legs	−.06	−.26	−.17	−.11	−.41*	.07
Feet	.07	.13	.22	.26	.03	.32*
Neck	.28	.05	.08	.35*	.10	.17
Torso	.02	.02	.12	.18	−.01	.37*
Sum	.06	.01	.12	.29	−.01	.39*

*$p < .05$

school or social adjustment in most of the subjects. Teachers seem more attuned to observe for acting out behavior.

From the parental ratings, a general picture emerges of children with constitutional short stature being socially withdrawn and aloof. They internalize emotional concerns as manifested by increased somatic complaints, schizoid tendencies, and social withdrawal. Some of the scores on the Child Behavior Checklist were so high that they reached the level typically found in children referred for psychiatric evaluation (Hollingsworth & Redlich, 1958). Our results are only partly consistent with prior anecdotal reports regarding the parental tendency to infantilize short children. Although the parents of our short children tended to set less clear behavioral limits and expected less obedience, they were not overprotective. It would be valuable to determine if these parents interact with the normal siblings in the same manner.

Although it is possible that parents of children with constitutional short stature are overly sensitized to the traits and experiences of their children, the responses of the subjects with constitutional short stature confirmed many of the observations of their parents. They viewed themselves less favorably than normal children. They had decreased self-esteem and more frequently described themselves as unhappy and unpopular with their peers. Such responses were independent of the age and sex of the child. Like the children with growth hormone deficiency (Drash, 1969; Money & Pollitt, 1966; Pollitt & Money, 1964), there were no significant differences between children with constitutional short stature and normal children on locus of control, sexual differentiation, and gender identity. Our results differ from those reported for children with growth hormone deficiency (Drotar et al., 1980; Stabler & Underwood, 1977), which found comparable psychological adjustment between affected children and their taller peers. Although the differences may be attributable to methodology, one must exercise caution when extrapolating observations from children with one form of short stature to another group, which might exhibit unique psychological characteristics.

Machover (1949) suggested that areas of conflict may be portrayed in drawings in ways in which the subject attempts to compensate for them. Therefore, we expected that the internalized concerns of our short children would be depicted in their drawings. Yet, we found no difference in the size, elaboration, or drawing sequence of the sexes between the two groups of children. Nor were there differences in four of six structural features of the drawings. Integration, transparencies, asymmetries, or slanting of figures did not indicate disturbances in self-concept, body image, or gender identity. Although short children used more shading in their drawings, the meaning of this with respect to their affect is unclear. Although excessive shading is generally indicative of anxiety (DiLeo, 1973; Jolles, 1964;

Machover, 1949), it does increase with the sophistication of the drawing (Swenson, 1968).

The most intriguing aspect of the drawings of these children was their propensity to omit body parts. Such omissions are considered to be among the most reliable signs for evaluating human figure drawings. Omissions in figure drawings are not particularly meaningful in children after age 10 due to their rarity (Vane & Eisen, 1962). Such findings in our short children may demonstrate immaturity, as our subjects included children up to age 13, with normal intelligence and academic abilities. Some authors view omissions of body parts as symbolic expressions of conflict (Jolles, 1964; Machover, 1949), but others (Koppitz, 1968; Stone & Ansbacher, 1965) propose that certain ones are associated with specific behavior problems. For instance, the omission of sensory organs is characteristic of shy, withdrawn, and depressed children (Koppitz, 1968; Stone & Ansbacher; 1965). We partially confirmed these findings in our short children. The omission of eyes correlated with increased schizoid tendencies and depression. There was only a trend toward their being non-communicative. Their tendency to omit extremities, unlike the omission of sensory organs, failed to confirm prior observations. Omissions of the lower extremities had previously been thought to reflect helplessness and insecurity (Koppitz, 1968; Michal-Smith & Morgenstern, 1969). In children with constitutional short stature, we found that omission of legs correlated with less aggression, and omission of feet, with hyperactivity. Likewise, we did not confirm the reported association of omission of parts of the upper extremities with increased aggression (Jolles, 1964; Koppitz, 1968). In summary, projective drawings have some value in discerning some personality problems of children with constitutional short stature. However, there is insufficient agreement among psychologists about the interpretation of many of the omissions.

The pattern of Rorschach scores also points toward a high level of introversion in children with constitutional short stature. The frequent use of human movement responses is typical of shy children, who usually cope with life situations by using ideation rather than action-oriented strategies. The increased frequency of morbid and oral-dependent content also suggests that our short children are generally more preoccupied with depressive thoughts and dependent longings. Their high barrier scores, although a positive sign of inner resources, also reflect the maintenance of firm boundaries between themselves and the outside world. These findings confirm parental descriptions of these children as being prone toward shyness and social withdrawal. They are in accordance with the children's responses of unhappiness and decreased social acceptance on the Piers-Harris Self-Concept Scale. Therefore, the personality style of children with constitutional short stature can be characterized as one of shyness and dependency.

The overall shyness of the children with constitutional short stature may refect the feelings and limitations that are imposed upon them by their short stature. They do not seem to be sufficiently confident to endure the hardships of social interactions and internalize life's problems. Perhaps, their introversion and dependency stem from the difficulties their parents have in treating them in an age-appropriate manner, thus inhibiting the child's sense of physical and social competence.

Our studies do not suggest other personality characteristics that have previously been attributed to short individuals. They are no more hostile, power-oriented, aggressive, preoccupied with sexual concerns, or less likely to establish affiliations than their normal peers. It is possible that in adulthood, some individuals who experienced constitutional delay over-compensate by extroversion or assertiveness. However, it is difficult to relate this to their prior short stature, because most attain normal adult height. We did not observe these traits in middle childhood.

We conclude that children with constitutional short stature share certain behavioral characteristics. They are comparable to the normal children across the majority of variables and have no significant arrest in personality development. Although occasionally there are short children who clearly suffer from serious psychological disorders, they are the exception rather than the rule. Psychologic adjustment does not hinge solely upon one's stature, but rather depends upon a multiplicity of factors, including family environment. Nevertheless, the aggregate of our data suggest that parents and teachers must become more attuned to the children's feelings of unhappiness and social withdrawal. Shy, quiet children often do not attract sufficient attention to their inner thoughts and feelings. All adults who intcract with short children must appreciate the importance of stature in our society, as well as how upset children become when they view their bodies as different. Children are unable to balance reassurance of future growth with immediate concerns about self-image. These children need the opportunity to discuss their feelings, receive acceptance and support, and be treated in accordance with their chronological age rather than their height age. Early intervention in childhood may prevent relatively minor psychological difficulties associated with their short stature from developing into major ones in adulthood.

ACKNOWLEDGMENTS

This study was supported in part by funds awarded to Robert A.Richman, Director of the Pediatric Endocrine Center, by the New York State Health Research Council, Contract 11-090, and the Human Growth Foundation. Additional support was received from the General Clinical Research Centers Program of the Division of

Research Resources, NIH grant RR229, awarded to the Clinical Research Center, State University Hospital, Upstate Medical Center.

The data were processed using a medically oriented data base management system, STATMANAGER (c) 1982, Hayden Software.

The authors would like to express their gratitude to Seymour Fisher for his guidance throughout the course of this project. The able assistance of Robert Kohlbrenner, Kathryn McCormick, Deborah Zimmerman, Eve Fisher, Pam Falter, and Joseph Pelligra is also greatly appreciated.

REFERENCES

Achenbach, T.M. (1979). The child behavior profile: An empirically based system for assessing children's behavioral problems and competencies. *International Journal of Mental Health, 7,* 24-42.

Achenbach, T. M., & Edelbrock, C. S. (1980). *Teacher form of the child behavior checklist.* Department of Child, Adolescent, and Community Psychiatry, University of Vermont.

Bender, L. (1938). A visual motor Gestalt test and its clinical use. *American Orthopsychiatric Association, #3.*

Conners, C. K. (1969). A teacher rating scale for use in drug studies with children. *American Journal of Psychiatry, 126,* 884-888.

Crain, W. C., & Smoke, L. (1981). Rorschach aggressive content in normal and problematic children. *Journal of Personality Assessment, 45*(1), 2-4.

DiLeo, J. (1973). *Young childrens' drawings as diagnostic aid.* New York: Brunner/Mazel.

Drash, P. W. (1969). Psychologic counseling in dwarfism. In L. I. Gardner (Ed.), *Endocrine and genetic diseases of childhood* (pp. 1014-22). Philadelphia: W. B. Saunders.

Drotar, D., Owens, R., & Gotthold, J. (1980). Personality adjustment of children and adolescents with hypopituitarism. *Child Psychiatry and Human Development, 11,* 59-66.

Dunn, L. M. & Markwardt, F. C. (1970). *Peabody Individual Achievement Test.* Minnesota: American Guidance Service.

Exner, J. E. (1974). *The Rorschach: A Comprehensive System I.* New York: Wiley.

Fisher, S., & Cleveland, S. (1968). Body image scoring system. In *Body image & personality* (2nd rev. ed., pp 395-414). New York: Dover.

Gold, R. F. (1978). Constitutional growth delay and learning problems. *Journal of Learning Disabilities, 11,* 36-38.

Goldfried, M., Stricker, G., & Weiner, S. (1971). Rorschach *Handbook of clinical & research Applications.* Englewood Cliffs, NJ: Prentice-Hall.

Gordon, M., Crouthamel, C., Post, E. M., & Richman, R. A. (1982). Psychosocial aspects of constitutional short stature: Social competence, behavior problems, self-esteem and family functioning. *Journal of Pediatrics, 101,* 477-480.

Gordon, M., Post, E. M., Crouthamel, C., & Richman, R. A. (1984). Do children with constitutional delay really have more learning problems? *Journal of Learning Disabilities, 17,* 291-293.

Greulich, W., & Pyle, S. I. (1959). *Radiographic atlas of skeletal development of the hand and wrist* (2nd ed.). Palo Alto, CA: Stanford University Press.

Haworth, M., & Normington, C. A. (1961). A sexual differentiation scale for the D-A-P Test. *Journal of Projective Techniques, 25,* 441-450.

Hollingshead, A. B., & Redlich, F. C. (1958). *Social class and mental illness: A community study.* New York: Wiley.

Horner, J. M., Thorsson, A. V., & Hintz, R. L. (1978). Growth decelerating patterns in children with constitutional short stature: An aid to diagnosis. *Journal of Pediatrics, 62,* 529-534.

Jolles, I. (1964). *A catalogue for the qualitative interpretation of the H-T-P (revised).* Beverly Hills, CA: Western Psychological Services.

Koppitz, E. M. (1963). *The Bender-Gestalt test for young children.* New York: Grune & Stratton.

Koppitz, E. M. (1968). *Psychological evaluation of children's human figure drawings.* New York: Grune & Stratton.

Machover, K. (1949). *Personality projection in the drawings of the human figure.* Springfield, IL: Thomas.

Masling, J., Rabie, L., & Blondheim, S. (1967). Obesity, level of aspiration and Rorschach and TAT measures of oral dependence. *Journal of Consulting Psychology, 31,* 233-239.

McArthur, R. G., & Fagan, J. E. (1971). An approach to solving problems of growth retardation in children and teenagers. *Canadian Medical Association Journal, 116,* 1012-1017.

McClelland, D. C., Clark, R. A., Roby, T. B. & Atkinson, J. W. (1949). The projective expression of need. IV. The effect of the need for achievement on thematic apperception. *Journal of Experimental Psychology, 39,* 242-255.

Michal-Smith, H., & Morgenstern, M. (1969). The use of the H-T-P with the mentally retarded child in a hospital clinic: In J. N. Buck & E. F. Hammer (Eds.), *Advances in the house-tree-person technique: Variations and applications.* Los Angeles: Western Psychological Services.

Money, J., & Pollitt, E. (1966). Studies in the psychology of dwarfism: II. Personality maturation and response to growth hormone treatment in hypopituitary dwarfism. *Journal of Pediatrics, 68,* 381-390.

Murray, H. (1943). *Thematic apperception test.* Cambridge: Harvard University Press.

National Center for Health Statistics. (1977). *Growth curves for children: Birth-18 years.* Hyattsville, MD: U.S. Department of Health, Education and Welfare.

Nowicki, S., & Strickland, B. R. (1973). A locus of control scale for children. *Journal of Consulting and Clinical Psychology, 40,* 148-155.

Piers, E. V. (1969). *Manual for the Piers-Harris Children's Self-Concept Scale.* Nashville: Counselor Recordings and Tests.

Pless, I. B., & Satterwhite, B. (1973). A measure of family functioning and its application. *Society for Scientific Medicine, 7,* 613-621.

Pollitt, E., & Money, J. (1964). Studies in the psychology of dwarfism. I. Intelligence quotient and school achievement. *Journal of Pediatrics, 64,* 415-421.

Pumroy, D. K. (1966). Maryland parent attitude survey: A research instrument with social desirability controlled. *Journal of Psychology, 64,* 73-78.

Shipley, J. E., & Veroff, J. (1952). A projective measure of the need for affiliation. *Journal of Experimental Psychology, 43,* 349-356.

Stabler, B., & Underwood, L. E. (1977). Anxiety and locus of control in hypopituitary dwarf children. *Research Relating to Children Bulletin, 38,* 75.

Stone, P. A., & Ansbacher, H. L. (1965). Social interest and performance on the Goodenough Draw-A-Man Test. *Journal of Individual Psychology, 21,* 178-186.

Swensen, C. H. (1968). Empirical evaluation of human figure drawings: 1957-1966. *Psychology Bulletin, 70,* 20-44.

Vane, J. R., & Eisen, V. W. (1962). The Goodenough Draw-A-Man Test and signs of maladjustment in kindergarten children. *Journal of Clinical Psychology, 18,* 276-279.

Veroff, J. (1957). Development and validation of a project measurement of power motivation. *Journal of Abnormal and Social Psychology, 54,* 1-8.

Wechsler, D. (1974). *Wechsler Intelligence Scale for Children—Revised.* New York: Psychological Corporation.

3 Effects of Short Stature on Social Competence

Deborah Young-Hyman
University of Maryland Medical School

The psychosocial literature concerning short-statured children is remarkable in that, with few exceptions, the conclusion is that shortness is a handicapping condition. This conclusion has led to an extensive effort by pediatric endocrinologists to seek effective forms of treatment, and has motivated mental health professionals to study these children. Three areas that have received a great deal of attention are personality characteristics, parent–child interaction, and cognitive and academic functioning. Positive adaptation might be operationalized as social competence. *Social behavior* is defined as that behavior that is culturally age appropriate, and competence as efficacy or skillfulness in accomplishing age-appropriate tasks. *Social competence* by this definition provides a measure of personality adjustment. Physical growth as a factor affecting psychosocial adjustment has not received a great deal of attention, and clarification of the relationship between growth and social competence seems indicated.

The salient characteristic of growth-delayed children and adolescents is that they are short, relative to peers, and that their stature increasingly differentiates them from age-mates as they grow older. This suggests that a developmental approach should be taken to assess their adaptation to short stature.

Studies of their personality characteristics have generally portrayed short children as psychologically maladjusted. In a study of children with constitutional short stature matched with normal stature controls, Gordon, Crouthamel, Post, and Richman (1982) concluded that "a general picture emerges of socially withdrawn and aloof children who express emotional concerns internally and tend to view themselves less favorably than do their taller peers" (p. 479). Steinhausen and Stahnke (1976) concluded that the group of growth hormone deficient children they studied were "living in a secluded inner world of intensified feelings, sentiments and emotions" (p. 782). These conclusions were based on personality tests that showed these children to be less tense, aggressive, excitable, shrewd and dominant, but more conscientious and tenderminded than short controls who were not growth hormone deficient. It is interesting to note, however, that some authors (Abbott, Rotnem, Genel, & Cohen, 1982; Rotnem, Genel, Hintz, & Cohen, 1977) comment that their subjects are able to perceive developmental tasks accurately, but feel incompetent and exhibit immature coping mechanisms.

The distinction between intrapsychic adjustment and accurate perception of socially appropriate behavior was also made in two independent studies by Drotar, Owens, & Gotthold (1980) and Stabler, Whitt, Moreault, D'Ercole, & Underwood (1980). Both of these studies reported that their primarily hypopituitary (all but three) short-stature subjects were able to perceive salient characteristics of sex role development, maturity of body image, "fun" activities and social situations, but coping skills brought to the solution of social problems and frustrating situations were less well developed. These authors conclude that the psychological adjustment of short-statured children compares favorably with normal-stature peers, *but* their social behavior tends to be less than adequate. Although supportive of the notion that short-stature children are in certain respects psychologically maladjusted, these findings underscore the necessity for making a distinction between the assessment of personality variables and the actual behavior exhibited in response to social situations.

Authors who looked at parent–child interaction and parent ratings of short children's behavior in an attempt to explain possible causes for social immaturity found that both parents and teachers reported a high incidence of immature and inadequate behaviors, and school adjustment problems. Patterns of maladaptive behavior were found to be etiology specific (Holmes, Hayford, & Thompson, 1982). These authors speculated that parental awareness of diagnosis, and possibly prognosis, may have contributed to parental attributions of poor adjustment. They cite as evidence that 25% of their sample had repeated a grade in school and that the retention was specifically attributed to size and perceptions of immaturity rather than academic achievement. In another study, parents who rated their short-statured children as having significant behavioral difficulties, somatic complaints, social withdrawal, and schizoid tendencies also reported poorer cooperation and communication among family members and a tendency to set less clear limits on their child's behavior (Gordon, et al., 1982). The tendency for parents to set inappropriate standards for the behavior of their short-statured child was also noted by Rotnem et al., (1977). These parents tended to set standards that were appropriate for a younger child and were overprotective and excessively controlling. The point is made, however, that not all children studied were maladjusted, and when families made efforts to foster a child's sense of competence, the child did not have significant developmental problems. This sentiment is echoed in the comment of Abbott et. al (1982) that "the diversity of adjustments of subjects in our studies appear to relate more to . . . the family's attitude toward a short stature child . . . than to the hypopituitarism itself" (p. 317).

Cognitive development and academic achievement have also been systematically studied in the short-statured population. There is general agreement that the IQs of short-statured children, regardless of etiology, approximate those of the general population and are independent of height (Meyer-Bahlberg, Feinman, MacGillivary, & Aceto, 1978; Money, Drash, & Lewis, 1976; Steinhausen & Stahnke, 1976). Academic achievement has been documented as being delayed but independent of cognitive abilities (Pollitt & Money, 1964). School adjustment problems and poor academic achievement have been attributed to emotional and/or size-related factors rather than the child's cognitive functioning (Holmes, et al. 1982). Evidence suggests that in the realms of both cognitive competence and parent--child interaction, factors other than literal competence or ability were contributing to the labeling of the child as maladjusted.

These findings call into question the validity of assumptions made about short stature. First, studies thus far that have found differing patterns of adjustment among etiology-specific short-statured children have not assessed height or degree of growth delay as an independent factor contributing to psychological adjustment. Second, the distinction between personality adjustment and social competence has not been made. Third, the validity of the assumption that short stature is a handicapping condition has not been tested. The present study was designed to assess these areas by posing the following questions: Are children who are significantly shorter than the general population socially incompetent? How does diagnosis, treatment, and change in stature affect social competence as reflected in the mastery of developmentally age-appropriate social tasks? Do differences exist between those children and adolescents who are ultimately found to have a diagnosable endocrine/medical disorder and those who are simply "short?" (i.e., constitutional short stature). This chapter describes the social competence of short-statured children and the attributions of competence made by themselves and their parents before etiology of their short stature is known. The relationships between growth delay, actual social competence, and attributions of competence are explored.

METHODS

Procedures.   Subjects were recruited from children and adolescents being evaluated for growth delay at University of Maryland Hospital, Division of Pediatric Endocrinology. In all cases the psychosocial protocol was administered to children and their parents before a diagnosis had been established. This was accomplished by testing subjects and their parents about the time of the child's admission to the Endocrine Diagnostic Unit for either a growth hormone stimulation test (Lanes & Hurtado, 1982; Raiti, Davis, & Blizzard, 1967) or a 24-hour integrated test of growth hormone production (Plotnick et al., 1975). Although some families knew the results of x-rays to determine the child's bone age (Greulich & Pyle, 1959), results of these tests were not known at the time of evaluation. Growth data including height, weight, growth rate, bone age, pubertal status, and parental height were available from the child's Endocrine clinic chart. Informed consent was obtained according to the provisions of the University of Maryland Medical School Human Volunteers Committee.

Interviewing and testing took place in such a manner that parent and child could not discuss the content of the evaluation materials before either responded. Children were always interviewed first, though parents were often present. Interview questions and questionnaires to which both parent and child responded were administered independently and/or simultaneously whenever possible. Children were specifically asked to respond for themselves and were encouraged to formulate their own opinions. Embellishments of children's responses by parents were not scored.

INSTRUMENTS

Structured Interview.   Each evaluation began with a structured interview. Information about the following areas was obtained: demographic information, family composition, grade and any specific class placement, health status of the child, mental health status of the child and the parent, parent and child perceptions of the source and nature of the child's growth problem, parent's and child's expectations of potential treatment and growth outcome, parent's and child's social activities and the child's peer relationships. To assess peer relationships, children were asked to report who their friends were. They were then asked names, sex, ages, length of relationship, and where they had met them, and to estimate the degree of intimacy. Social activities were assessed by asking the children what things they did besides attend school. Their responses were coded into social activity categories that closely approximate those of the Social Competence Scale of the Achenbach Child Behavior Checklist (Achenbach, 1979). Activities were counted if they were structured, took place at a regular time, and could be corroborated by the parent. For example, receiving lessons on a musical instrument was counted as a sedentary activity, as was playing a piano, if the parent reported the child did this activity on a regular basis. Peer relationships were also corroborated by parent report but parental embellishments to either question were not scored. Coding of the responses to interview questions was cross-validated by an independent rater for a random sample of the subjects. Interrater reliability was approximately 80% and a consensus score obtained.

Silhouette Apperception Test (Grew, Stabler, Williams, & Underwood, 1983).   Using sexually undifferentiated drawings that correspond to the 3rd, 10th, 25th, 50th, 75th and 97th percentiles on the growth chart (Hamill, et al., 1979), the respondent is first asked to match the child's stature to the one that most closely corresponds to his or her height in comparison to same-aged peers. The respondent is then instructed to match the child's stature "when they were all through growing and all through with whatever (if any) treatment the doctor might give them" in comparison to fully grown men and women (as was appropriate) as represented by the figures. This task was administered to the child and to the parent.

Wechsler Intelligence Scale for Children-Revised (Wechsler, 1974).   The entire WISC–R was administered with the exception of the Digit Span and Mazes subtests. The purpose was to obtain a current assessment of the child's intellectual abilities.

Bender Visual-Motor Gestalt Test (Bender, 1946).   This is a test of visual-motor integration skills. The Koppitz Developmental Scoring System (Koppitz, 1963) was used to obtain an age equivalence. All Bender tests were rescored by an independent rater and a consensus score obtained if a discrepancy existed.

Perceived Self-Competence Scale (Harter, 1982).   This instrument was developed to assess attributions of competence in cognitive, social, and physical skills, and to obtain an independent measure of self-esteem or self-worth. Questionnaire items were selected so that statistically independent factors were obtained in each skill domain assessed. Statements are presented in a forced choice format and the respondent is asked to compare the child's behavior to that of his or her peers. The child form is written in the generic "Some children" format and the teacher form contains equivalent questions that begin "This kid." In lieu of a parent form, parents filled out the teacher form with the instruction "Read these statements as if they said "My child." Normative data are available in the form of mean scale scores by grade.

Means–Ends Problem Solving Test (Platt, Spivack, & Bloom, 1971). Children are asked to make up the middle of a story about the solution of a social-interpersonal problem. The beginning and the end of the story are provided by the examiner with the instruction to make the middle of the story as complete as possible. Story roots are divided into those appropriate for children 12 years and below, and those appropriate for adolescents and adults. Each story is scored for number of means, obstacles and expressions of time used to achieve the stated end of the story. Scoring guidelines are presented by the authors. Three age-appropriate stories were presented to each child and the average number of the aforementioned factors utilized in achieving the stated story end were computed for each subject. This instrument has been shown to discriminate between psychiatrically and behaviorally disturbed versus normal populations. Some normative data are available in the form of mean scores by age/grade and socioeconomic status (SES) (Shure & Spivak, 1972). Scoring was cross-checked by an independent rater and consensus scoring obtained.

SUBJECTS

Twenty-seven subjects and their parents were ultimately entered into the study. The sample is clearly skewed in the direction of upper-middle-class families. The modal family was in the professional/major business class according to the most recent Hollingshead Scale (1975). Subjects' age ranged from 8 years 8 months to 15 years 10 months. Twenty-one (78%) were males, and 6 (22%) were females. Males and females did not differ on demographic characteristics, but females tended to be younger (see Table 3.1). Children with a chronic health or mental disability, including a previously diagnosed health or learning problem for which they were currently receiving some form of remediation or treatment, were excluded from the study.

RESULTS

The girls were smaller and lighter on measurements of height, height age ($t = 4.03, p < .001$ and $t = 3.79, p < .001$ respectively) and weight, ($t = 2.88$, $p < .01$) (Table 3.2). Relative smaller stature would be predicted given the girl's younger chronological age and the expected differences in height and weight between males and females of comparable age.

TABLE 3.1
Demographic Information on 27 Study Subjects

N		Males 21	Females 6
Mean Chronological Age		$12^{1}/_{12} \pm 2^{1}/_{12}$ yr	$10^{5}/_{12} \pm 1^{4}/_{12}$ yr
SES[a]	Mode	5	5
	Range	3 (19%)	3 (17%)
		4 (33%)	4 (33%)
		5 (48%)	5 (50%)
Grade Placement	Mode	7th	4th and 7th
	Range	3rd–10th	4th–7th
Family Composition	Intact	17 (81%)	4 (67%)
	Single Parent	2 (9.5%)	1 (16.5%)
	Step-Parent	2 (9.5%)	1 (16.5%)

[a]Hollingshead Four Factor Scale of Social Status, 1975.

TABLE 3.2
Growth Data

		Males	Females
Mean Height Age		$9^{3}/_{12} \pm 2$ yr	$7^{4}/_{12} \pm {}^{5}/_{12}$ yr
Mean Visible Growth Delay[a]		$2^{11}/_{12} \pm 1^{1}/_{12}$ yr	$3^{2}/_{12} \pm {}^{9}/_{12}$ yr
Bone Age		$10^{2}/_{12} \pm 2$ yr	$8^{2}/_{12} \pm 2$ yr
Bone Growth Delay		$2^{2}/_{12} \pm 1^{1}/_{12}$ yr	$2^{2}/_{12} \pm 1$ yr
Mean Weight Percentile		5.2 ± 5.7	7.0 ± 8.8
Pubertal Status[b]	Prepubertal	16 (76%)	4 (67%)
	Tanner 2	4 (19%)	2 (33%)
	Tanner 3	1 (5%)	
Age of Onset of Growth Delay		4.1 yr	3.2 yr
Parental Height Percentile	Mean	38 ± 42	11 ± 13
	Range	3rd to 93rd	3rd to 28th
Diagnosis	Growth Hormone Deficient	10 (48%)	2 (33%)
	Constitutional Delay of Growth	10 (48%)	2 (33%)
	Genetic Short Stature	3 (14%)	1 (17%)

[a]Visible growth delay is chronological age minus height age.
[b]Pubertal status was determined by using the rating of the child's most advanced state of sexual development according to Tanner criterion.

Assignment to diagnostic category was made by the criterion listed in Table 3.3. Twelve of the children (44%) were diagnosed as Growth Hormone Deficient, 11 (41%) as Constitutional Delay of Growth. The remaining 4 children (15%) had no abnormal medical test findings but were significantly below the 3rd percentile in height relative to chronologic age, and were diagnosed Genetic Short Stature.

The results of the children's performance on tests of cognitive abilities, perceived competence, and social problem solving are summarized in Table 3.4. Mean Full Scale IQ for all subjects fell within the bright normal range of intelligence (110), ranging from low normal (85) to very superior (142). This is not an unexpected finding given the high level of education of the parents. Perceptual-motor skills as measured by the Bender–Gestalt test were significantly delayed ($\bar{x}$ age equivalence 9 years 2 months) relative to chronological age ($\bar{x}$ chronological age 11 years 8 months). Only one child performed at age expectancy.

Children rated themselves as significantly more competent than age-norms in the areas of cognitive abilities ($t = 2.57, p < .02$), social skills ($t = 4.0, p < .001$), and self-esteem ($t = 4.56, p < .001$) on the task of Perceived Competence. Self-ratings in the domain of physical competence were not significantly different from normative data ($t = 0.6$, N.S.). Parents' and children's ratings of the child's competence were significantly different in the areas of cognitive functioning and self-esteem. Children rated themselves significantly lower in cognitive competence ($t = -4.34, p < .001$) and marginally lower in self-esteem ($t = -1.75, p < .09$). Children's ability

TABLE 3.3
Diagnostic Criterion

Growth Hormone Deficient	- Peak GH of < 10 ng/ml after administration of a provocative stimulus (ICTT or AITT) and/or
	- 24-hour average concentration of growth hormone < 3.2 ng/ml and
	- Bone age delay ≥ 2 SDs from chronological age and
	- Growth rate ≤ 4.2 cm/yr
Constitutional Delay of Growth	- Bone age delay ≥ 2 SDs from chronological age and
	- 3rd percentile height[a] or below
Genetic Short Stature	- 3rd percentile height[b] or below
	- Bone age within 2 SDs of age expectancy

[a]One child was below the 9th percentile but had a bone age delay > 2 SDs from chronological age.

[b]One child was on the 5th percentile for height.

TABLE 3.4
Psychological Test Data

		Sample	Males	Females
WISC-R Full	x̄:	110 ± 16	111± 17	110 ± 13
Scale IQ	Range:	85 to 142	85 to 142	96 to 126
Bender Visual-Motor Gestalt	x̄:	9 yrs. 2 mos.	9 yrs. 4 mos.	8 yrs. 6 mos.
Test Age Equivalence		± 1 yr. 2 mos.	± 1 yr. 2 mos.	± 10 mos.
Perceived Competence Scale Child-Cognitive	x̄	3.02 ± .64[a]	3.01 ± .72	3.05 ± .26
- Social	x̄:	3.20 ± .59	3.20 ± .57	3.20 ± .72
- Physical	x̄:	2.80 ± .64	2.80 ± .64	2.7 ± .67
- Self-Esteem	x̄:	3.17 ± .44[b]	3.2 ± .47	3.1 ± .34
Parent-Cognitive	x̄:	3.39 ± .50[a]	3.3 ± .53	3.6 ± .33
- Social	x̄:	3.26 ± .74	3.20 ± .75	3.50 ± .72
- Physical	x̄:	2.89 ± .74	3.00 ± .70	2.60 ± .82
- Self-Esteem	x̄:	3.37 ± .56[b]	3.30 ± .61	3.6 ± .25
Silhouette Apperception Test Child Now	Mode:	3rd (52%)	3rd (48%)	3rd (67%)
	Range:	3rd to 75th	3rd to 75th	3rd to 25th
Parent Now	Mode:	3rd (63%)	3rd (57%)	3rd (83%)
	Range:	3rd to 10th	3rd to 10th	3rd to 10th
Means-Ends Problem Solving Test	x̄:	2.69 ± 1.85	2.79 ± 1.74	2.33 ± 2.34

[a]Difference between parent's and child's rating, $t = -3.60, p < .001$.
[b]Difference between parent's and child's rating, $t = -1.75, p < .09$.

to solve social problems (mean total solutions = 2.69) were lower than the mean number of solutions ($\bar{x}$ = 4.39) in a sample of similarly aged middle-class children reported by Shure and Spivack (1972).

When subjects and their parents were asked to assess the child's height relative to same-aged peers, both children and their parents were quite accurate in their perceptions. The most frequent response was the 3rd percentile figure, with 52% of the children and 63% of the parents responding in this manner. Only five children in the sample saw themselves as larger than the 10th percentile compared to peers. None of the parents rated their child above the 10th percentile.

Peer relationships and social activities are reported in Table 3.5. These children present themselves as a socially active group: mean total social activities per child = 3.67, with an average of 3.3 close friends. Parents affirm this by reporting 81% of their children to be as popular as, or more popular than, age-mates (see Table 3.10). Only one child reported no close

TABLE 3.5
Peer Relationships and Social Activities

		Sample		Males		Females	
Peer Relationships:							
Number of Friends	x̄:	3.3		3.2		3.7	
	Range:	0 to 6		0 to 6		2 to 6	
Sex of Friends							
	Same Only:	16	(59%)	13	(62%)	3	(50%)
	Opposite Only:	0	(0%)	0	(0%)	0	(0%)
	Both:	11	(41%)	8	(38%)	3	(50%)
Age of Friends							
	Younger:	7	(25%)	7	(33%)	0	(0%)
	Same or Older:	20	(74%)	14	(67%)	6	(100%)
Intimacy	Close:	14	(52%)	11	(52%)	3	(50%)
	Casual:	1	(4%)	1	(5%)	0	(0%)
	Both:	12	(44%)	9	(43%)	3	(50%)
Structure of Contact	One to One:	14	(52%)	11	(53%)	3	(50%)
	Group:	3	(11%)	3	(14%)	0	(0%)
	Both:	10	(37%)	7	(33%)	3	(50%)
Length	1 Year:	3	(11%)	3	(14%)	0	(0%)
	1 Year:	24	(89%)	18	(86%)	6	(100%)
Social Activities:							
Total	x̄:	3.67	(100%)[a]	3.24	(100%)[a]	5.17	(100%)[a]
Sports	x̄:	1.30	(74%)[a]	1.33	(71%)[a]	.67	(67%)[a]
Religious	x̄:	1.07	(70%)[a]	.91	(62%)[a]	1.33	(83%)[a]
Intellectual/							
Sedentary	x̄:	.96	(48%)[a]	.81	(48%)[a]	1.67	(100%)[a]
Activity Group	x̄:	.22	(22%)[a]	.10	(9%)[a]	1.5	(50%)[a]

[a]Percentage of children reporting one or more activities in this category.

friends, and none reported only friends of the opposite sex. An inverse relationship was found, however, between the age of the child and the number of friends reported ($r = -.5796, p < .001$), as well as the tendency to have a close friend of the opposite sex ($r = -.2777, p < .08$). Subjects described themselves as having primarily long-term friendships (over 1 year in length) and almost half (44%) described themselves as having both close and casual friends. As a group, subjects reported that they preferred to interact with friends on a one-to-one, as opposed to group, basis (52% vs. 37% respectively). The most frequently reported group activity was, however, organized sports. Of the subjects, 74% participated in one or more league sports at the time of the evaluation, but age tended to be negatively correlated with participation in sports activities ($r = -.2961, p < .07$). The next most frequent extracurricular activity reported was attendance at one or more regular religious functions (70% of all subjects). No relationship was found between age and this activity. Total number of extracurricular activities did, however, decrease as the children became older ($r = -.3408$, $p < .04$).

Relationships between growth parameters and reported social activities are reported on Tables 3.6 and 3.7. Number of friends was found to be inversely related to actual height ($r = -.4942$, $p < .004$), height age equivalent ($r = -.4943$, $p < .004$) and a measurement of visible growth delay (chronologic age minus height age, $r = -.3417$, $p < .04$), bone age ($r = -.4493$, $p < .009$), pubertal status ($r = -.5153$, $p < .003$), and age of onset of short stature ($r = -3357$, $p < .04$). The formation of a close

TABLE 3.6
Relationships Between Growth Parameters and
Reported Social Activities: Entire Sample

	Religious	Intellectual/ Sedentary	Activity Groups	Total Extracurricular Activities
Age			$r = -.3238**$	$r = -.3408**$
Height			$r = -.3873**$	
Height Age			$r = -.3765**$	
Visible Growth Delay	$r = -.3518**$			
Perceived Growth Delay		$r = .3232**$		
Weight			$r = -.3137**$	
Diagnosis	$r = .4250*$	$r = -.3238**$		

Significance Levels of Pearson r's, $*p < .01$. $**p < .05$.

TABLE 3.7
Relationships Between Growth Parameters
and Peers/Friendships: Entire Sample

	Number of Friends	Length of Relationships
Age	$r = -.5796*$	
Height	$r = -.4942**$	
Height Age	$r = -.4943**$	
Visible Growth Delay	$r = -.3518****$	$r = .4128****$
Weight	$r = -.4303***$	
Pubertal Status	$r = -.5153**$	
Age Onset Short Stature	$r = -.3357****$	$r = -.3903****$
Bone Age	$r = -.4493***$	$r = .3538****$

Significance Levels of Pearson r's, $*p < .001$. $**p < .005$. $***p < .01$. $****p < .05$.

friendship with a person of the opposite sex tended to be more likely the shorter the child was relative to chronological age ($r = -.2667, p < .09$) and the earlier the onset of short stature ($r = -.2858, p < .07$). Length of children's relationships was also related to lower age of onset of short stature ($r = -.3903, p < .02$), severity of growth delay as measured by bone age ($r = -.3538, p < .05$), and visible growth delay ($r = .4128, p < .02$). The more long-standing and greater the growth delay, the longer children maintained friendships.

Measures of actual height, height age, weight, and bone age were found to be inversely related to participation in activity groups (e.g., Cub Scouts) ($r = -.3873, p < .02, r = -.3765, p < .03, r = -.3137, p < .05$ and $r = -.2981, p < .06$ respectively), as was age ($r = -.3238, p < .05$). Children who were more advanced in pubertal development and perceived themselves to have a greater growth delay reported more intellectual and sedentary activities ($r = .3069, p < .06$ and $r = .3232, p < .05$ respectively). Diagnostic category was related to the pursuit of intellectual and sedentary activities, and age of friends. Children with fewer medical findings reported fewer intellectual activities ($r = -.3238, p < .05$) and more same-aged or older friends ($r = -.5581, p < .001$). These same children reported higher participation in religious functions ($r = -.4250, p < .01$). The only growth parameter related to total number of social activities was bone age ($r = -.3606, p < .03$). The greater the bone age delay, the more social activities a child reported.

Growth parameters and their relationships with parent/child ratings of competence are reported in Table 3.8. Children who rated themselves as more cognitively able and as having higher self-esteem were taller ($r = .3115$ and $r = .3103$ respectively, $p < .06$), heavier ($r = .2901, p < .07$ and $r = .1.3240, p < .05$ respectively) and more well developed sexually ($r = .4773, p < .006$ and $r = .5097, p < .003$ respectively). Parents' ratings of cognitive competence were similarly related to the child's height ($r = .3565, p < .03$) and pubertal status ($r = .3964, p < .02$). Parental attributions of the child's level of self-esteem was also positively related to pubertal status ($r = .3045, p < .06$) but negatively related to parental height ($r = -.3932, p < .02$).

The negative relationship between parental height and attributions of competence was true in the social sphere for both parents and their children ($r = -.4446, p < .01$ and $r = -.3376, p < .04$ respectively). Taller parents perceived their children to have lower self-esteem and poorer social skills. Parents also saw their children as less socially competent the taller the children were ($r = -.3003, p < .006$). This is probably the result of the larger growth delay in the older children. Age of onset of short stature was the only growth parameter found to be significantly associated with physical competence. Children who had been growth delayed longer

TABLE 3.8

Relationships Between Growth Parameters and Child/Parent Ratings of Competence: Entire Sample

	Cognitive, Parent	Social, Child	Physical, Child	Physical, Parent	Self-Esteem, Child
Height Age	r = .3365***				
Visible Growth Delay		r = .3291***			
Weight	r = .4044***				r = .3240***
Pubertal Status	r = .4773**				r = .5097*
Age Onset Short Stature			r = −.4782**		
Parental Height Percentile		r = −.3376***		r = .4446**	

Significance Levels of Pearson *r*'s, *p* < .005. **p < .01. ***p < .05.

perceived themselves to be more physically competent ($r = -.4782$, $p < .006$). Despite the fact that neither parents nor children knew their diagnosis, both rated those children with fewer medical findings as more cognitively competent ($r = .3324$, $p < .05$ and $r = .3023$, $p < .06$ respectively).

As can be seen in Tables 3.9 and 3.10, parents associated participation in sports with social and physical competence as well as the child's self-esteem and popularity ($r = .4640$, $p < .007$; $r = 3216$, $p < .05$; $r = .3457$, $p < .04$ and $r = .3718$, $p < .03$ respectively). The strong association between participation in sports and physical competence may be due to the sports-related criterion respondents are asked to use to rate physical competence. Parents' ratings of their child's social competence and popularity are also strongly related to the presence of a close friend of the opposite sex ($r = .4799$, $p < .006$ and $r = .5174$, $p < .003$ respectively), the complexity of their relationships, i.e., both close and casual friends ($r = .4825$, $p < .05$ and $r = .5309$, $p < .002$ respectively) and individual and group contexts ($r = .5470$, $p < .002$ and $r = .5324$, $p < .002$ respectively). Parents' independent ratings of their child's popularity correspond closely with the child's report of number of close friends ($r = .4109$, $p < .02$). In contrast, parents' attributions of cognitive competence were strongly related to the child's participation in intellectual or other sedentary activities ($r = .4786$, $p < .006$). Children also perceived their social and physical competence to be related to participation in sports ($r = .3220$, $p < .05$ and $r = .4341$, $p < .01$

TABLE 3.9
Relationships Between Perceived Competencies and Reported Social
Activities: Entire Sample

	Sports
Perceived Competence Scale—Child	
Cognitive	
Social	$r = .3220$**
Physical	$r = .4341$*
Self-Esteem	
Perceived Competence Scale—Parent	
Cognitive	
Social	$r = .4640$*
Physical	$r = .3216$**
Self-Esteem	$r = .3457$**
Parent Rating of Child's Popularity	$r = .3718$**
Loner = 5 (19.5%)	
Average = 10 (37%)	
Very Popular = 12 (44%)	

Significance Levels of Pearson r's, *$p < .01$.   **$p < .05$.

TABLE 3.10

Relationship Between Perceived Competencies and Reported Peers/Friendships: Entire Sample

	Number of Friends	Sex of Friends	Degree of Intimacy	Structure	Length of Relationships
Perceived Competence Scale—Child					
Cognitive	$r = .3422***$	$r = .3533***$	$r = .4862*$	$r = .5386*$	$r = .3273***$
Social		$r = .3439***$			$r = .3806***$
Physical					
Self-Esteem					
Perceived Competence Scale—Parent					
Cognitive					
Social		$r = .4799**$	$r = .4825*$	$r = .5470*$	
Physical					
Self-Esteem					
Parent Rating of Child's Popularity	$r = .4109***$	$r = .5174*$	$r = .4309**$	$r = .5324*$	

Significance Levels of Pearson r's, $*p < .005$. $**p < .01$. $***p < .05$.

respectively). Their positive association between cognitive competence and intellectual and sedentary activities ($r = .3687, p < .03$) was similar to their parents'. Children's perception of their cognitive competence was also associated with more extracurricular activities ($r = .3594, p < .03$) and higher verbal abilities ($r = .5434, p < .002$). Higher verbal abilities were also associated with higher self-esteem ($r = .3665, p < .03$). Higher verbal and performance IQ scores predicted more total number of social activities reported ($r = .3660, p < .03$ and $r = .3239, p < .05$ respectively). Attribution of physical competence was negatively related to performance IQ scores ($r = -.3237, p < .05$); however, the physical competence scale of the Perceived Competence Scale assesses different skills than the performance skills required on the WISC-R.

DISCUSSION

Results of this study suggest that short stature alone does not represent a handicapping condition, but that age of onset of growth delay and the degree of both the perceived and real growth delay may be more critical factors affecting social competence.Family environment and other compensatory factors may mediate the social behavior of short children.

Children and their parents in this study do not represent the general population. They are clearly more socially advantaged, as evidenced by their higher socioeconomic status and the children's higher than average IQs. The supportive nature of the family environment is underscored by the fact that these were families who were coming to the medical community for the sole purpose of evaluation and potential treatment of short stature. Many parents stated that they wished to give their children every chance to reach both their maximal social and growth potential. This optimistic and healthy attitude was also reflected in the lack of psychopathology in the parents. Other authors (Kusalic & Fortin, 1975) have found a high percentage of the families they studied to be pathological and note "covertly rejecting parental attitudes towards the dwarfed patients" (p. 328). A significant number of the parents in this study accurately perceived their children's popularity and cognitive ability, were supportive, and gave their children credit for social competence. It is noteworthy that parents who are short themselves see their children as more socially competent. The importance of a supportive family environment to the psychological adjustment of short-statured children has been found by a number of other authors (Abbott et al., 1982; Rotnem et al., 1977).

An important finding in this study was that cognitive abilities were valued independently of height and physical competence. Previous studies have suggested that parents' and teachers' views of cognitive ability are strongly influenced by size (Holmes et al; 1982) and smaller stature may significantly affect school achievement (Pollitt & Money, 1964). Family membership in a socioeconomic class in which cognitive abilities are highly valued may have facilitated the children's shift from sports activities, in which they were considered less competent, to intellectual pursuits, in which they excelled. They were therefore able to maintain both a high level of self-esteem and social competence.

Other compensatory social strategies suggested by this study are children's preference for long-standing friendships and complex relationships. Short stature that becomes evident in early childhood seems less critical to the formation of peer relationships, especially those with friends of the opposite sex. Children with early onset short stature formed appropriate peer relationships early and maintained them over long periods of time. Many reported having the same set of friends from age 2 or 3. This is in contrast to studies that have reported short-statured children to have a high degree of social isolation (Drash, Greenberg, & Money, 1968) and significant difficulty with heterosexual relationships (Lewis, Money, & Bobrow, 1973). Children also spoke of preferring to interact on a one-to-one basis, which permitted them to use their highly developed verbal skills.

The necessity of distinguishing between how short-statured children perceive themselves and their actual involvement in peer and social activities was confirmed. Older, more severely growth-delayed children perceived themselves to be less socially and physically competent. This supports findings from other studies that report children to have feelings of alienation from their peer group (Rotnem, Cohen, Hintz, & Genel, 1979) as well as lower self-esteem (Gordon et al., 1982). However, no unusual patterns of peer relationships were reported by the sample represented in this study. Lack of a comparison group of normal-statured age-mates prevents assessment of whether the decrease, with age, in sports and overall number of extracurricular activities is developmentally normal. The validity of comparing hypopituitary short-statured children, who have other potentially handicapping conditions, and those with other growth disorders must also be questioned.

Findings of the present study have implications regarding both medical and psychological treatment of children who are short. Although an analysis of participation in social activities and peer relationships by specific age categories was not possible given the small sample size, the relationship between age, decreasing perceived social competence and participation in social activities needs further exploration.

Timing of both medical and/or psychological intervention may be critical to long-term psychological adjustment as reflected by participation in age-appropriate activities. The focus of the intervention should also be carefully considered. This study affirms the conclusion of other authors, which has been that familial environment may be an important contributory factor to the social adjustment of short-stature children (Drotar et al., 1980). Additionally, because satisfactory growth outcome is not guaranteed with treatment (Benjamin, Muyskens, & Saenger, 1984; Grew et al., 1983) and children have been documented to become depressed as a result of perceived medical treatment failure (Rotnem et al., 1979), psychological intervention might best be focused on family attitudes and compensatory areas of mastery.

REFERENCES

Abbott, D., Rotnem, D., Genel, M., & Cohen, D. J. (1982). Cognitive and emotional functioning in hypopituitary short-statured children. *Schizophrenia Bulletin, 8,* 310–319.

Achenbach, T. M. (1979). The child behavior profile: An empiracally based system for assessing children's behavioral problems and competencies. *International Journal of Mental Health, 7,* 24–42.

Bender, L. (1946). *Bender Visual-Motor Gestalt Test: Cards and manual of instructions.* New York: American Orthopsychiatric Association.

Benjamin, M., Muyskens, J., & Saenger, P. (1984). Short children, anxious parents: Is growth hormone the answer? *Hastings Center Report, 14* (2), 5–9.

Drash, P. W., Greenberg, W. E., & Money, J. (1968). Intelligence and personality in four syndromes of dwarfism. In D. B. Cheek (Ed.), *Human growth: Body composition, cell growth, energy and intelligence.* Philadelphia: Lea & Febiger.

Drotar, D., Owens, R., & Gotthold, J. (1980). Personality adjustment of children and adolescents with hypopituitarism. *Child Psychiatry and Human Development, 11* (1), 59–66.

Gordon, M., Crouthamel, C., Post, E. M., & Richman, R. A. (1982). Psychosocial aspects of constitutional short stature: Social competence, behavior problems, self-esteem, and family functioning. *Journal of Pediatrics, 101,* 477–480.

Greulich, W. W., & Pyle, S. I. (1959). *Radiographic atlas of skeletal development of hand and wrist* (2nd ed.). Stanford: Stanford University Press.

Grew, R. S., Stabler, B., Williams, R. W., & Underwood, L. E. (1983). Facilitating patient understanding in the treatment of growth delay. *Clinical Pediatrics, 22,* 685–690.

Hamill, P. V. V., Drizd, T. A., Johnson, C. L., Reed, R. B., Roche, A. F., & Moore, W. M. (1979). Physical growth: National Center for Health Statistics percentiles. *American Journal of Clinical Nutrition, 32,* 607–629.

Harter, S. (1982). The perceived competence scale for children. *Child Development, 53,* 87–97.

Hollingshead, A. B. (1975). *Four Factor Index of Social Status.* New Haven: Yale University.

Holmes, C. S., Hayford, J. T., & Thompson, R. G. (1982). Parents' and teachers' differing views of short children's behavior. *Child Care, Health and Development, 8,* 327–336.

Koppitz, E. M. (1963). *The Bender-Gestalt test for young children.* New York: Grune & Stratton.

Kusalic, M., & Fortin, C. (1975). Growth hormone treatment in hypopituitary dwarfs: Longitudinal psychological effects. *Canadian Psychiatric Association Journal, 20,* 325–331.

Lanes, R., & Hurtado, E. (1982). Oral clonidine — an effective growth hormone-releasing agent in prepubertal subjects. *Journal of Pediatrics, 100,* 710.

Lewis, V. G., Money, J., & Bobrow, N. A. (1973). Psychologic study of boys with short stature, retarded osseous growth, and normal age of pubertal onset. *Adolescence, 8,* 445–454.

Meyer-Bahlburg, H. F. L., Feinman, J. A., MacGillivary, M. H., & Aceto, T. (1978) Growth hormone deficiency, brain development, and intelligence. *American Journal of Disease of Children, 132,* 565–572.

Money, J., Drash, P. W., & Lewis, V. G. (1967). Dwarfism and hypopituitarism: Statural retardation without mental retardation. *American Journal of Mental Deficiency, 72* (1), 122–126.

Platt, J. J., Spivack, G., & Bloom, M. (1971). *Means-ends problem-solving procedure (MEPS): Manual and tentative norms.* Philadelphia, PA: Department of Mental Health Sciences, Hahnemann Medical College and Hospital.

Plotnick, L. P., Thompson, R. G., Kowarski, A. A., deLacerda, L., Migeon, C. J. & Blizzard, R. M. (1975). Circadian variation of integrated concentration of growth hormone in children and adults. *Journal of Clinical Endocrinology and Metabolism, 40,* 240–247.

Pollitt, E., & Money, J. (1964). Studies in the psychology of dwarfism. I. Intelligence quotient and school achievement. *Journal of Pediatrics, 64,* 415–421.

Raiti, S., Davis, W. T., & Blizzard, R. M. (1967). A comparison of the effects of insulin hypoglycemia and arginine infusion on release of human growth hormone. *Lancet, 2,* 1182–1183.

Rotnem, D., Cohen, D. J., Hintz, R., & Genel, M. (1979). Psychological sequelae of relative "treatment failure" for children receiving human growth hormone replacement. *Journal of the American Academy of Child Psychiatry,* 18, 505–520.

Rotnem, D., Genel, M., Hintz, R. L., & Cohen, D. J. (1977). Personality development in children with growth hormone deficiency. *Journal American Academy of Child Psychiatry, 16,* 412–426.

Shure, M. B., & Spivack, G. (1972). Means-ends thinking, adjustment, and social class among elementary-school-aged children. *Journal of the Consulting and Clinical Psychology, 38* (3), 348–353.

Stabler, B., Whitt, J. K., Moreault, D. M., D'Ercole, A. J., & Underwood, L. (1980). Social judgements by children of short stature. *Psychological Reports, 46,* 743–746.

Steinhausen, H. C., & Stahnke, N. (1976). Psychoendocrinological studies in dwarfed children and adolescents. *Archives of Disease in Childhood, 51,* 778–783.

Wechsler, D. (1974). *Wechsler Intelligence Scale for Children-Revised.* New York: Psychological Corporation.

4 Psychosocial Aspects of Short Stature: The Day to Day Context*

Leonard P. Sawisch
Michigan Department of Education

It is an honor to be invited to keynote this event and I want to thank the Human Growth Foundation and Serono Symposia for that honor. I have been assured as keynote speaker, that there would be no timing buzzer. I've been asked, however, to be sensitive to any synchronized snoring from the audience. In that regard, I have good news and bad news. The good news is I was able to condense my talk down to 21 slides and charts; the bad news is it's locked in my trunk in Lansing, Michigan.

It's one of those inevitable things—as I was getting out of the car at the airport somebody walked up and said, "Well, you're kind of short." Every time that happens, it catches me by surprise and I don't know what to say. So this time I just looked at myself and said, "Wow, and it must have just happened!" Then I walked off without my slides!

I have to admit that when this happens, it tends to insult my sense of pride. But I think it is also important for me to be honest and say I wasn't always proud to be a dwarf. In fact, it wasn't that many years ago that I first came out of the closet as a dwarf. Actually it was a clothes hamper, but the point is, I made a very important self-realization. I realized, "hey, maybe people weren't avoiding me because I was a dwarf—maybe people were avoiding me because I was an ass." My point, quite simply put, is that we cannot hope to change the situation we find ourselves in until we begin to take some responsibility for that situation, and realize that we as individ-

---
*Dr. Sawisch's contribution to this volume is taken from his keynote address at the Psychosocial Aspects of Growth Delay Conference, October 19-21, 1984.

uals play a significant role in what goes on around us. Based on this observation, I would like to share some things with you this evening.

The title of my presentation is the "Psychosocial Aspects of Short Stature." My job is to give you a context to help you understand some of the discussions we've heard in the last two days about growth delay, which is part of the short stature experience. I would like to do this in what I call the "day-to-day" context. What I'm talking about, of course, is the "dwarf experience"—I'm not talking about a medical problem; I'm not talking about a physical condition; I'm talking about a *sociocultural* phenomenon. Before I move on, I would highly recommend Joan Abilon's book, *Little People in America, the Social Dimensions of Dwarfism.* It's an excellent reference, which covers many of the points that I will be sharing with you this evening.

To illustrate my thesis, I want to cover four areas. First, I want to talk to you about how society tends to view little people. In order to do that effectively, I need to share with you how society tends to view handicappers as a group, with little people being part of that group. Then, I want to talk about how we, as little people, parents of little people, and handicappers in general, tend to view ourselves in response to the social view. After that, I want to share with you a different perspective, a different way of understanding that reaction. And finally, we'll talk about some of the implications this has for handicappers in general, for individuals with delay growth, for their parents, and for practitioners.

SOCIETY'S VIEW

How does society tend to view handicappers? I first became interested in this issue when, as a young child, I was watching television and saw the March of Dimes campaign commercial. I'm sure many of you remember this: "Help wipe out birth defects in your lifetime." Being a somewhat precocious child, it seemed interesting and probably the noble thing to do, so I remember sending off for the brochures. Six weeks later, when the mail arrived with the literature from the March of Dimes, you can imagine my surprise when I read that I was listed as a birth defect. Now, being a precocious child, putting two and two together I realized, "hey, it may not be that hard to wipe out birth defects in my lifetime."

This was a very frightening concept and I had never thought about it that way. On one hand we are spending thousands and thousands of dollars every year as a society to incorporate individuals who are considered to be handicappers, people like me who are to be mainstreamed into society. On the other hand, we're spending thousands and thousands of dollars every year to avoid people like me being born. There are, in fact, segments of society that suggest that if born, handicappers should not be allowed to live. That's confusing to me, because I wonder which way it is? On one hand, I'm supposed to feel encouraged to participate fully in society, but on the other hand, I'm supposed to be encouraged not be be here!

It seemed to me that what I was experiencing was a double message; so I began to look around for other double messages. I didn't have to look very far. I remember first getting turned on to the labels that society feels comfortable using to describe people like us. We're known as "the handicapped," "the disabled," "the impaired," "the deformed," "the dysfunctive," "the defective," "the in-val'id,"—excuse me, that's "in'valid," I slipped on that one. All these labels have a tremendously negative valence to them. Most people would feel very uncomfortable labeling themselves in these ways, yet we are expected to accept these labels without question. All of this added to my confusion.

I listened to the little bits of advice that society would tend to give people, like, "Everyone has a handicap." Now, you've heard that, "Everyone has a handicap." If everyone has a handicap, why do they call us "the handicapped?" I couldn't understand that. Everyone has a handicap. If everyone has a handicap, then the concept is obviously irrelevant, and yet the fact that we are here today suggests that it's very relevant. It's confusing.

"You are only handicapped if you think you are." You've heard that, right? You are only handicapped if you think you are. You want access to appropriate education? Well then, we must document the fact that you are handicapped. You are only handicapped if you think you are. You want access to appropriate vocational training? Then we must document the fact that you are handicapped. You are only handicapped if you think you are. If we want access to the city bus, some of us must document the fact that we are handicapped; yet we are only handicapped if we think we are. You know from your own personal experience that the agencies in our home towns have thick files documenting the fact that we are handicapped, and yet we are only handicapped if we think we are. It's very confusing.

You begin to understand that there is no middle ground, that somehow, as handicappers and as little people, we're expected to either be super-cripples or basket cases. Somehow there is no room in the middle. For example, if we do something good or expected or even average, people are surprised. "You drive?" "You date?" On the other hand, if we do something bad or inappropriate or below average, of course, it's because we're dwarfs; it's because we're handicapped! It's one way or the other; there is no middle ground.

It's very interesting if we look at mixed marriages. I'm talking about a handicapper married to non-handicapper, or in this context, a littler person married to an average-statured person. It's interesting to think about the social attributions that go on there. We look at the dwarf in that situation and say, "My, hasn't he done well for himself." Then we look at the non-dwarf in that situation and say, "Why is he throwing his life away?" This is a very interesting double message.

I found it refreshing that the researchers reporting at this Symposium were very honest in expressing their surprise that we're not as psychologically, socially, or academically screwed up as they originally thought. Of course, as some folks did suggest, it was probably an *error* in their data! But the point is, we are not allowed, we're not expected, to be just average; and this is the pervasive social message.

Think about the children's literature that your parents read to you and perhaps you have read to your children. Think about how we depict good and evil. We use handicappers as a symbol of super good or super evil. If you want to make somebody evil in literature, what do you do? Chop off a hand, poke out an eye, Captain Hook; chop off a leg, Moby Dick. We use handicapper characteristics to embody evil; like Rumplestiltskin, that gnarled little dwarf who steals your firstborn child. (I've always wondered what he did with all those children!)

On the other hand, we have super good. Jolly Old St. Nicholas, right? The Jolly Old Elf is the embodiment of goodness. There is an interesting corollary phenomenon that goes on here. As St. Nick becomes more popular, he gets bigger and bigger, right? Of course, his workers are still dwarfs and we assume they are working at minimum wage!

My favorite is Snow White and the Seven Dwarfs. Growing up in a town where I was the only little person, you can imagine that book had a profound impact on me. When all my friends were out dating, I was out looking for six other dwarfs so I could date, too! It never dawned on me that they weren't particularly good role models. I mean, here you have seven guys living out in the woods alone. They have (this is an adult crowd) one woman among them. She's still a virgin, and she runs off and marries someone else. These guys are out in the mines every day, doing hard physical labor with picks and shovels, and what are they like? They're whistling! All of the time, they're whistling! It finally dawned on me—they were on drugs! In the same story, when the Queen goes through her ritual to embody evil, what happens? She gets old and decrepit and develops arthritis. Super good—super evil.

You see the same thing in television. Remember when "Ironside" first came on? It was supposed to be such a positive image for handicappers in general. Here's a person using a wheelchair in the hilliest city in the country, right? He gets into every building. He was, of course, celibate. He occasionally had a lunch or dinner date, but that's where he drew the line. The most unrealistic part was that nobody stared; nobody ever pointed. It was just not real. One of my other favorites—"Ze plane, ze plane." A pet dwarf on national TV! We see the same in movies. The "Joni" movie—we had a whole series of those kinds of movies—"I broke my neck and found God." Or the travesty of "Under the Rainbow."

This depiction is very interesting, because when you do see handicappers on TV or in the movies, they're invariably playing the role of handicappers. You don't see us depicted as the shopkeeper down the street or the neighbor next door. Again, the message is that you're not supposed to like who or what you are if you're a dwarf, if you're a handicapper.

SELF-VIEW

The question becomes: if that is the way society tends to view us, how do we tend to view ourselves? I think that the most obvious way of describing that is "confusion." I grew up, like most little people, appearing to be very precocious and received a lot of positive attention. At a very young age, I figured I was a pretty cool dude and I tended to like myself. Then it became more and more obvious that I was a dwarf, which of course is a bad thing to be, and it became confusing. How could I be a good person and be a dwarf, which is a bad thing? If I was a dwarf I couldn't be a good person. If I was a good person I couldn't be a dwarf. It's a very interesting struggle between those two perspectives that we don't often talk about.

It's also interesting because we, in a clinical setting, see people struggling with this kind of thing and we wonder why it doesn't appear in some of our data. I think part of the reason is that the measures we use for self-esteem are unidimensional and don't allow us to deal with the fact that someone can feel good about themselves in one part and not feel good about themselves in another part. In a clinical setting, we are likely to see little people as "little people." Little people's lives, however, are not spent as little people but as brothers or sisters, as schoolchildren or something else. Therefore, for most of their lives that conflict does not exist. But when it is there, it is very confusing and can be very painful.

Those of you who are parents are familiar with this struggle, because you went through it long before your children. That's something we often overlook—that this little identity crisis, if you will, is something that parents confront first. It's one of the reasons that some little people find it difficult to be with other little people. If a part of you says, "I'm a good person," the last person you want to see is another dwarf, because that reminds you of the fact that you're a dwarf (which is a bad thing to be). As a dwarf, you can't be a good person, and then you are confused again.

We see this not only with little people but with handicappers in general, and society is not totally insensitive to that. Society has assumed that it is important for people to feel good about themselves, so we offer them one of two major options. The first is that we encourage people to emphasize the positive and de-emphasize the negative. "Sure, you're short, fat, and ugly—now don't think about it." It works very well, right? It's interesting because we've heard that kind of advice given but we rarely stop to think, "What does it mean?" It means, "accept that there is something wrong, accept that there is something bad about you, and then don't think about it."

For many of us, this identity crisis can be resurrected in the form of the "dislocated shoulder syndrome." That's when you're in the store and the little kid says, "Mommy, Mommy, look at the. . . . " She jerks his arm out of the socket, and we have another child with a misconception about little people. It was funny because I was in the store the other day, and whenever this happens I try to make some visual contact with the kid involved (without their parents seeing it) to let the kid know it was okay. I was in the process of attempting to do that when my wife grabbed my arm and jerked me around and said, "Come on, it's not polite to stare."

The effort to emphasize the positive and de-emphasize the negative starts by telling you to accept that there is something wrong with you. The other advice we give, and perhaps this is more common, is "Be thankful." We have a young woman here who has spina bifida and is paralyzed from the waist down. How do we help her feel good about herself? Assuming that we want to do that (most of us feel a need to do that) what we tell her is, "Be thankful, be thankful that you can see and that you can hear and that you're not mentally retarded." Hey, this is very interesting, because it has the very same underlying message. On the streets it goes like this, "You may be shit, but there are people out there who are shittier than you." That's how you feel good about yourself! This is a very interesting form of condoned prejudice that says, "Accept that there is something wrong with you, and then find somebody who's *worse* and dump on them." We may not think about that when we hear this advice, but the underlying message is there.

Now, if you'll look at the political situation of adult handicappers in our country today, you see the effects of this message because there is a tremendous amount of ranking that exists. When you talk to deaf separatists, their message is basically, "Deafness is a damned inconvenience; we just thank God we're not blind, crippled, or mentally retarded." When you talk to the blind separatists you hear, "Blindness is a damned inconvenience; we just thank God we're not deaf, crippled, or mentally retarded." I'm a member of Little People of America (kind of makes you want to pick us up and hug us, doesn't it?). Our motto is, "Little people think big." Sure, and Black Panthers think white. I understand. It makes perfect sense to me. But if you listen, what's our message? "We're just little people; thank God we're not blind, deaf, crippled, or mentally retarded!" You see, we've all bought into the idea that we're somehow inferior, and we have therefore been encouraged to pull rank on other handicapped people.

The point is that we see dwarfism as an inferior state of being. That came out in one of the presentations this morning when someone said that it is difficult to get a control group because of the "moral and ethical issues.", i.e., it's immoral and unethical to allow someone to be a dwarf if you can do anything about it! This means, for us, that in order *to feel good we have to accept that we are bad.*

A DIFFERENT APPROACH

Now as you might guess, I couldn't accept that.There had to be an alternative; there had to be a different approach. I feel somewhat like Luther because the inspiration came to me on the privy. I'm 4'4" tall, and my wife is 3'9" tall. We were sitting around a number of years ago thinking, "That's our bathroom in there. Wouldn't it be nice if we could use it?" So, for her birthday, I built the floor of the bathroom up so that the toilet seat was no more than 9" from the floor. Now hold that thought in your mind for a minute.

When I go out in public, to the mall, and I feel the need to use the public room, I go in and balance myself precariously on the stool to do my business. When I'm done, I come out, wash my hands, reach up for a towel and the water runs down my arm! Then I have to climb up into the wet sink to reach the mirror to comb my hair. Society looks at me in that situation and says, "Isn't it a shame? Isn't it too bad that God did that or that happened?" Society sees me experiencing a problem and blames me, assuming that there is something wrong with me. So, you know what I do? I invite those people to come to my house to use my bathroom, please! It's great. The guys are in there experiencing vertigo. People come out and their legs are all cramped up, and, of course, I love it. I stand outside and say, "My God, you're handicapped, you're disabled." Invariably the response is, "Bullshit, it's the toilet."

That's when it all came clear to me. That's how I finally understood the double standard. When I or another little person or a handicapper experiences a problem, we've been encouraged to blame it on ourselves. We've been encouraged to think that there is something wrong with us, with our emotions, with our minds, our bodies. Yet when other people experience the same kinds of problems, they are encouraged to blame it on the environment, and their solutions, of course, focus on the environment. It finally dawned on me that I too could blame the toilet. That I too could look at the problem as the interaction between me and my environment, and not have to accept that there was something wrong with me.

When you think about it, that's essentially the same kind of consciousness-raising process that most minority groups in this country have experienced. For example, women come to realize, "Yes, I'm different from a man, but he's just as different from me. The difference goes both ways. There's nothing wrong with being different. What is wrong is that we put a value judgment on being different." That is what is wrong; that's where the pain comes from. Difference does not mean less than. Therefore, I said to myself, "If it's working for these groups, why not for me, why not for us?"

It wasn't an easy thing to do. I began playing with the question, "Well, what's wrong?" I think we as parents hear this all the time: "What's wrong with your kid?" "I don't know, but don't touch it." People come up and invade your space with that question, "What's wrong?" So I began to ask myself, "What's wrong? What's wrong with being a dwarf? What's wrong with being a handicapper?"

As I was going through this process I was at Michigan State University, and I met a man who had cerebral palsy. In the first conversation, he said, "I have a speech defect, so if you don't understand what I'm saying let me know and I will repeat it." And I said, "Speech defect? Why, I noticed you had an accent, but why you call it a defect is beyond me." This was a tremendously enlightening experience for him, because no one had ever put it to him quite that way. But why not?

Around the same time, I saw a woman on campus outside the journalism building. She used a wheelchair and she was obviously distraught. I approached her and said, "You really seem to be bummed out (because that's the way we talked back then)—what's the problem?" She said, "Well, I'm paralyzed from the waist down, I can't get into the journalism building, I need a class there to graduate, and I'm not going to be able to graduate. I don't know what I'm going to do." I said, "Wait a minute. If we took those steps away and put a ramp there, could you take the class?" "Well, yeah." I said, "Then why in hell are you blaming your body when the problem is the building?" She had never thought of it that way.

Let's talk about the bottom line. What's wrong with being mentally retarded? You are more likely to be physically attacked, you are more likely to be sexually abused, you are more likely to be ripped off in the marketplace, you are more likely to be denied appropriate education, you are more likely to be denied appropriate vocational training, you are more likely to be denied a job, you are more likely to be denied an opportunity to live in society with other people, and you are more likely to be socially ostracized in almost every setting. 'That's what's wrong with being mentally retarded. It has almost nothing to do with mental retardation per se.

IMPLICATIONS

This was a very, very interesting coming-out process for me.The implication was that I should obviously begin to feel pride in being a dwarf. It seemed to be the next logical step, that there was nothing wrong with it, and yet I remember when the thought first hit me, my stomach tied up in a knot. I mean, literally. Proud to be a dwarf? How could that possibly be? I was keeping a journal at the time and it took me 3 months to get comfortable with the concept of being proud of being a dwarf.

Of course, at the end of 3 months, I was really into it. I'm talking total immersion. I was playing drums at the time and I billed myself as Dewey Dwarf on Drums. On St. Patrick's Day I came dressed all in green as Larry Leprechaun. One day I came in all dressed in black with a white tie—Melvin Midget out of Chicago. I went so far as to get a white Afro wig and came in as Peter Pygmy, Albino. I was totally into it. I had "Dwarf Power" sewn on the back of my jacket, and "Dwarf Power" embroidered on my sleeve. The point is, I had spent 20 some years of my life ashamed . . . ashamed of being a dwarf. It was important for me to spend some time being intensely proud so that I could get it in balance; so that I would neither be ashamed or excessively proud to be a dwarf; so I could find that middle road.

When you begin to develop a sense of pride in yourself, the next natural thing is to begin to develop a sense of pride in your people. Now for me, that coming-out process meant identifying with other little people and that's when I found out that Attila the Hun was a dwarf. Aesop was a dwarf, a black, hunchbacked dwarf. The electric filament for the light bulb was perfected by an immigrant hunchbacked dwarf, and a deaf guy named Edison got credit for it. (I never understood that.) Tom Thumb, a real estate tycoon, worked in the sideshows for just a few years and sold out for whole bunches of money.

Look at the contemporary situation—Linda Hunt, the Oscar winning actress; Doug Herland, the Olympic medal winner; Joe Coonha, the world ranked power lifter; Pat Delray, the publicist; Lee Kitchens, the inventor and entrepreneur; Gary Coleman, the celebrity. The list just goes on and on. Yet I had never been encouraged to make that identification, and I had never been encouraged to take pride in the fact that I was a dwarf, in spite of the fact that dwarfs have made some significant contributions.

The more I thought about it, the more I realized that Little People of America, Inc., one of the oldest self-help groups in this country, is one of the few handicapper groups that not only provides support to its members (little people) but also to their parents. I think one of the most fascinating parts about it is that, as far as I can tell, it is the only handicapper self-help group that formed outside of a medical context. The other self-help groups formed around growing medical interest, but for us it was the other way around. In fact, it was our support and, many times, our tolerance that allowed for the significant growth and interest in genetic and orthopedic issues. It is clearly something that we as a people can feel pride in.

Once you make the identification with other dwarfs, it's hard not to identify with other handicappers. I was surprised to find out that no one ever told me that three of our greatest presidents were handicappers; that the greatest nuclear scientist of our age has muscular dystrophy, and that the greatest violinist of our age is post-polio; that Rockefeller and Leonardo da Vinci were learning disabled, and that Einstein was considered stupid until he got out of school; that Homer was blind and Beethoven was deaf. I wondered why no one had stressed this in school, or why handicapper children had not been turned on to their culture and their heritage.

When you take pride in yourself, you take pride in your people. You realize that those whom you see as your people keep growing. I guess this is the best part of all. . . . When you take pride in yourself and pride in your people, whomever you define as your people, it's almost impossible not to make that next step, the most important step, and that is taking pride in all people. That, it seems to me, is the bottom line, the real message.

The irony of this concept, relevant to what we have been studying here, is that growth-delayed individuals who may eventually reach "average size" need to accept that being a dwarf is okay, in order to feel that being average sized is also okay. Parents may also need to come to this same conclusion, in order to find the peace that they seek. For practitioners it suggests that we still need to address the *real* moral and ethical issue in relation to the use of growth hormones.

SUMMARY AND CONCLUSIONS

We have talked about how society views handicappers as inferior, and how we as handicappers tend to view ourselves in response to this social view. I have shared with you my concept of a different approach—if you remember nothing else, remember my toilet. I talked a little bit about the progression of pride, from pride in yourself to pride in all people. For me, this is the crux of understanding the day-to-day context of short stature.

One of the things that I think we've learned at this particular Symposium is an appreciation of how people are similar. It's by appreciating how people are similar that we get to the point of really enjoying the differences between people. As I was struggling with this concept a number of years ago, I was the father of a 1-year-old and a 3-year-old, and as most parents do, I would interject this into the conversation whenever I could with a certain amount of pride. Invariably someone would say, "Are your kids small, too?" And I'd respond, "So far!"

I had a hard time trying to figure out what they expected so I'd go on, "Yes, I have a 3-year-old who has already outgrown my clothes, but we're close! When I read him a story at night, he sits me on his lap. And when he falls asleep, I carry him to bed, and his little feet drag on the floor."

That's when Lee Kitchens set me straight. He said, "The only problem you're going to have with your kids is potty training." First off, potty training is a euphemism. You don't have to train your potty, the darn thing just sits there. It's the kid you've got to train! You've got to convince them that what they have been carrying around in their diapers for the last couple years is no longer cool. No more of that cute, "I smell someone!" You start getting straight with them, "(Sniff) is that you???" Then you have to use logic, good old American logic, to convince them to flush it down the toilet—into our water system. We taxpayers then pay millions of dollars each year to clean it out of there; and to top it off, we go down to Frank's Nursery and buy it back in a bag and spread it on our lawn! That's when I realized I would be like Lee and cut out the middle man; I just send my kids outside!

The point is we all live in a world that's just a little bit crazy. One of the most exciting things, however, is that from what I can tell there has never been a better place or a better time in the history of this world to be a little person than right now, today. And yet if we pool our talents, our concerns, and our time, and if we promote our own sense of independence and if we celebrate our rich culture and heritage, then the times have no choice but to be even better.

5

The Relationship of Academic Achievement and the Intellectual Functioning and Affective Conditions of Hypopituitary Children

Patricia Torrisi Siegel
Children's Hospital of Michigan
Detroit, Michigan

Nancy J. Hopwood
C. S. Mott Children's Hospital
Ann Arbor, Michigan

Hypopituitarism affects approximately 15,000 children in the United States. These children have slow growth due to growth hormone deficiency, either secondary to insufficient hypothalamic stimulation of the pituitary or pituitary gland malfunction (Owen & Root, 1979). Growth, although slow, is proportionate and there are no deforming or unusual features. Intelligence is not thought to be affected, but academic achievement is often poor. Affectively, hypopituitary children are described as having low self-concept resulting from the consequences of short stature and over-protective, restrictive child-rearing practices. An investigation of the relationship of academic achievement to the intellectual functioning and affective conditions of hypopituitary children constitutes the primary focus of this study.

Poor academic achievement in hypopituitary dwarfs has been explained in the literature according to three conflicting theories: low achievement as a function of cognitive underfunctioning—low self-concept; as commensurate with subaverage ability; and as secondary to specific cognitive atypicalities.

Cognitive underfunctioning is the most frequently cited explanation of poor academic achievement in hypopituitary children. According to this theory, failure to achieve up to average intellectual potential is said to be secondary to low self-concept. Low self-concept, in turn, is said to be a consequence of short stature and specific child-rearing practices (Kusalic & Fortin, 1975; Money & Pollitt, 1966; Rotnem, Genel, Hintz, & Cohen, 1977). A *psychomaturational lag* or emotional immaturity in these youngsters was first described by Money and Pollitt (1966). The authors report "the more infantilized the youngster, the greater the immaturity in emotional development." Poor academic achievement is viewed as one manifestation of this emotional immaturity.

A second, less frequently cited theory addressing the relationship between ability and achievement in hypopituitary youngsters states that poor school performance is commensurate with subaverage ability (Obuchowski, Zienkiewiez, & Graczykowska-Koczorowska, 1970; Spector, Faigenbaum, Sturman, & Hoffman, 1979). Obuchowski et al. reported intellectual dullness in 15 Polish hypopituitary adults. Poor academic achievement, inferred because many had not completed high school, was interpreted as consistent with depressed ability. Similarly, in a study of 20 short-statured females, Spector et al. (1979) reported that the 4 with hypopituitarism had low IQs relative to elevated socioeconomic status (SES). Achievement scores were below grade level, and interpreted by the authors as commensurate with low ability.

The theory that poor academic achievement in hypopituitary children can be explained relative to specific cognitive deficits has not been empirically tested. However, it is reasonable to assume that some of these children are similar to those frequently described as "learning disabled" for two basic reasons. First, both reportedly have overall average ability, but do poorly in school. Second, both have been characterized as having central nervous system (CNS) immaturity secondary to significant birth and neonatal histories. For example, it is well recognized that children considered high risk for learning problems frequently have traumatic birth histories and/or complications during the neonatal period (Clements, 1966; Kaevi & Pasamanick, 1958). Developmental delays are also frequently reported (Kinsbourne, 1973; Ross, 1976), as well as CNS dysfunction (Critchley, 1970; Dykman, Ackerman, Clements, & Peters, 1971). In a study investigating the incidence of perinatal insult in 46 children with idiopathic hypopituitarism, Craft, Underwood, and Van Wyk (1980) found that 65% had histories of at least one risk factor such as significant gestational bleeding, prematurity, breech delivery, or evidence of fetal distress (asphyxia) at birth. Similarly, Kusalic and Fortin (1975) found that most of the 11 hypopituitary children in their sample had "long delivery and often breech presentation." They also reported a slight lag in all aspects of development. Finally, Meyer-Bahlburg, Feinman, MacGillivary, & Aceto (1978) found that 9 of 28 subjects comprising their sample had a history of brain dysfunction or neurological abnormalities, providing another similarity between learning disabled and hypopituitary children. It remains to be determined, however, whether other characteristics typical of children with learning deficits are also present in hypopituitary children. Specifically, if hypopituitary children exhibit cognitive variability in terms of verbal-performance skills and visual-motor integration skills that frequently characterize the learning profiles of children with learning problems, do they have specific learning deficits?

The present study investigates the relationship of cognitive and affective factors to the academic achievement of hypopituitary children relative to the three theoretical explanations proposed. Clarification of the relationship may have significant implications for the development of appropriate educational programs for hypopituitary children. Many of these youngsters experience grade retention because of poor school performance and because they look or act immaturely. Further clarification of their learning characteristics and behavioral status may encourage the development of strategies other than retention for meeting their educational needs.

SAMPLE SELECTION AND DESCRIPTION

The parents of 55 children between the ages of 4 and 16 currently treated for idiopathic hypopituitarism by endocrinologists at C.S. Mott Children's Hospital in Ann Arbor, Children's Hospital of Michigan (CHM) in Detroit, and Children's Hospital in Buffalo, New York, were sent a letter describing the project. Fifty-three parents (96%) agreed to have their child participate. All 53 children were administered the psychometric battery, but 11 were excluded from the final sample. Four were excluded because subsequent diagnostic procedures indicated craniopharyngioma or deprivation dwarfism. Six children were excluded because their chronologic age (<6 years) precluded adequate achievement testing. Finally, one child was excluded because severe mental retardation (IQ <30) precluded the use of the psychological instruments selected for the study. The 42 children in the final sample were school aged and definitively diagnosed as having idiopathic hypopituitarism according to the following medical criteria established by the National Hormone and Pituitary Program, Baltimore. These criteria are:

1. growth hormone secretion of less than 7 ng/dl in response to two provocative stimuli, arginine and insulin
2. a delayed bone age
3. growth of less than 4 cm over the past year
4. height below two standard deviations from the mean for age, or below the 3rd percentile.

Demographics for the total hypopituitary sample are presented in Table 5.1. According to traditional endocrinological standards, 28 of the growth hormone deficient (GHD) children have isolated growth hormone deficiency (IGHD) and 14 have multiple hormone deficiencies (MHD). Children were classified as MHD if they had GHD and also were currently receiving oral replacement medication for adrenocorticotropic hormone (ACTH) and/or thyroid stimulating hormone (TSH) deficiency. Deficiency in luteinizing hormone (LH) or follicle stimulating hormone (FSH) was not included in the criteria for the MHD classification because most of the children were too young for definitive gonadotropin assessment. Specifically, 13 have TSH deficiency, with 8 having both TSH and ACTH deficiency. Among the 13 children with TSH deficiency, 10 began replacement therapy subsequent to growth hormone replacement; 3 were deficient at the time growth hormone deficiency was diagnosed. Only 1 of these 3 was less than 4 years of age.

The sex ratio shows the expected preponderance of male children with 30 (71%) boys and 12 (29%) girls. Sex comparisons on the various age measures revealed no significant differences. The Hollingshead Two-Factor Index of Social Position was used to measure social class background and was compared to the distribution of employed persons in Michigan according to the 1970 Michigan Census Data (Table 24, General Social and Economic Characteristics). A Chi Square Goodness of Fit Test revealed no significant difference between the sample and the general Michigan population (df = 4) (^{2} = 4.35 NS). Chronologic age (CA), bone age (BA) at IQ testing, age diagnosed GHD (Age Dx), age growth hormone replacement began (Age Rx), and duration of treatment with replacement therapy are also presented in Table 5.1. All but 7 were regularly receiving growth hormone replacement at the time of IQ testing. These 7 began therapy within 3 months.

Developmental histories including medical risk factors and milestone attainment are presented in Table 5.2. Data gathered from the official medical records of 38 children revealed a high incidence of birth trauma

TABLE 5.1
Demographics for the 42 Hypopituitary Children

		Hollingshead Index for Social Position			
	Low	La	Ave	Ha	High
Number of children	4	14	13	9	2
Age Measures		Range		$\overline{X}$	SD
Chronologic Age		6^5–16^6		12^2	2^4
Bone Age		3^5–12^3		8^7	2^5
Diagnosed		6 mos–13^0		6^6	2^9
Treatment Initiated		2^5–13^0		7^0	2^4
Duration of Treatment		0–8^6		3^5	3^3
School Grade		1–10		5	1^5

Note.   Superscripts indicate months.

TABLE 5.2
Developmental Histories for the Hypopituitary Children

Medical History Risk Factors[a]		
	N	%
Prematurity	12	32%
Breech	4	11%
Anoxia	6	16%
Seizures	10	26%
Hypoglycemic Seizures	4	11%
At Least One Risk Factor	20	53%

Developmental Milestones Attainment					
Walked[b]			Talked[c]		
Mos	N	%	Mos	N	%
< 12	7	18%	12–17	13	38%
12–17	21	55%	18–20	10	29%
≥ 18	10	26%	> 21	11	32%

[a]Medical History was not available for four children.
[b]Four mothers did not answer the question.
[c]Eight mothers did not answer the question.

and/or neonatal distress. Twelve of the hypopituitary children were born prematurely (32%) compared to an incidence of 10% in the general population (Craft et al., 1980). Similarly, 4 were breech presentation (11%), an occurrence rate three times greater than the 3.5% incidence of the general population (Craft et al., 1980). In addition, 6 children (16%) had significant postpartum distress (anoxia), whereas 10 children (26%) had seizures in the neonatal period. In at least 4 of these infants, however, the seizures were known to be secondary to hypoglycemia. A total of 20 children (53%) had at least one significant risk factor in their neonatal medical history. Sex comparisons on risk factors revealed no significant differences.

Information regarding developmental milestone attainment was collected from a parent questionnaire. Of the GHD children, 10 (26%) did not begin walking independently until 18 months or later. Similarly, 11 (32%) did not begin using two-word phrases until 21 months or later. There were no sex differences noted. Although wide variation in attainment of developmental milestones in normal children is acknowledged, most children begin to walk and use two-word sentence structures by 15 months, and 18 months respectively (Mussen, Conger, & Kagan, 1974).

Information regarding school histories was also collected from the parent questionnaire. Current grade in school at IQ testing ranged from first grade to tenth grade with a mean grade placement of fifth grade. Sixteen children (38%) have failed at least one grade in school. Sex comparisons on school history data revealed no significant differences.

The IGHD and MHD groups were compared on all demographic and developmental measures. The two groups were found to be proportionate relative to sex distribution with the IGHD group having 20 (71%) males and the MHD group having 10 (71%) males. There were no significant differences between the two classifications on social position or age measures. The MHD group had a higher incidence (36%) of clinical seizures than did the IGHD group (18%). This is not an unexpected finding and may reflect the increased probability of seizure activity in children with MHD secondary to hypoglycemia (Hopwood, Forsman, Kenny, & Drash, 1975). The incidence of developmental delay was also higher for the MHD children. Specifically, 36% of the MHD children did not walk alone until 18 months or later compared to 18% of the IGHD children. Similarly, 36% of the MHD children did not begin to use two-word phrases until after 21 months of age compared to 21% of the IGHD children.

METHODS AND PROCEDURES

Each youngster was administered a battery of psychometric tests by a certified psychological examiner to assess cognitive functioning (intelligence and visual-motor integration), affective functioning (self-concept), and academic achievement (reading and math).The mother of each child was given a self-administered parent attitude inventory.

Intelligence was assessed by the Wechsler Intelligence Scale for Children-Revised (WISC-R). The Verbal IQ (VIQ), Performance IQ (PIQ) and Full Scale IQ (FSIQ) were derived for all children. An absolute Verbal-Performance (V-P) difference score was derived for each child. The magnitude of the V-P discrepancy scores were grouped according to "frequency of occurrence" in the normative standardization sample (Kaufman, 1975). An absolute V-P discrepancy score of 18 points occurs only 1 out of 15 times in the general population and is considered abnormal (Kaufman, 1979).

Visual-motor integration was assessed by the Bender Gestalt Test. The Koppitz (1963) scoring system was used to derive a developmental level of visual-motor integration by converting the number of developmental errors to a percentile score using the conversion table provided in the test manual. Because maturation is known to affect visual-motor development and because there is a question of developmental delay among hypopituitary children, a brain injury indicator score (Koppitz, 1963) was also derived from each Bender protocol. A child's Bender performance was considered indicative of a significant visual-motor integration deficit if the developmental score was < 16th percentile, *and* four or more errors were significant indicators of brain injury.

The Piers–Harris Self-Concept Inventory (P-H) was administered to 38 of the children. Three children were excluded because CA (<8 years) was below standardization. An additional child's inventory was invalidated secondary to low intellectual functioning. A general self-concept raw score was derived according to procedures described in the manual.

Parent attitudes were assessed by the Parent Attitude Research Instrument (PARI). The PARI is a self-administered inventory consisting of 23 five-item scales designed to be used in a variety of situations where parent attitudes toward child-rearing and family life are to be related to parent--child relationships and the personality development of children (Schaefer & Bell, 1958). Five relatively independent factors have been identified. Two of these are used in the present study: *Harsh-Punitive Control,* which reveals an overdominance by the mother, and *Overpossessiveness,* which reveals a covert control of the child through keeping the child indebted to the mother, dependent and immature. In scoring the PARI, weights of 4, 3, 2, and 1 are assigned to the response categories of strong agreement to strong disagreement, where a score of 1 is reflective of a healthier attitude, a score of 4 a pathogenic attitude. Factors scores were derived by calculating the mean of the scales comprising the factor. Mean raw scores were converted to stanines using the table provided by the authors of the instrument.

Academic achievement was assessed using two subtests of the Wide Range Achievement Test (WRAT): Reading Sight Word Recognition and Math and one subtest of the Peabody Individualized Achievement Test (PIAT): Reading Comprehension. Standard scores were derived using the conversion tables in the manuals. The two standard scores for reading were averaged to obtain a single achievement level in reading. Low academic achievement was defined by a standard score <85 in one or both achievement measures. Children identified as low achievers were grouped according to the support their individual psychometric profiles provided for one of the three theories explaining academic failure:

1. *Cognitive deficit theory*—at least one WISC-R scale score falls within the average range (90–110); a V-P difference >18 points and/or a significant visual-motor integration deficit.

2. *Low ability theory*—both WISC-R scale scores fall below the average range (<90).

3. *Cognitive underfunctioning—low self-concept theory*—both WISC-R scale scores fall within the average range; there is neither a significant V-P difference nor a significant visual-motor deficit.

RESULTS

Mean VIQ and PIQ scores for the hypopituitary sample are significantly lower than mean scores for the normative sample but fall within the average range (Table 5.3). The mean absolute V–P difference score for the hypopituitary children is significantly higher than that of the normative sample. More importantly, almost twice the percentage of sample children (29%) had V–P differences of a magnitude > 18 points, compared to 16% in the normative sample (Kaufman, 1979) (Fig. 5.1).

The mean general self-concept score for the sample is significantly higher than the mean of the standardization sample (Table 5.3). This difference is not likely to be of any practical significance because means reported in several studies of large groups of school-aged children subsequent to standardization have been similar (Michael, Smith, & Michael, 1975). However, it is surprising given the widely accepted belief that hypopituitary children have low self-concepts. It was suspected that self-concept may be influenced by the age at which growth hormone replacement therapy was initiated and/or by the duration of replacement therapy. Correlations between self-concept and initiation of treatment ($r = -.076$) and duration of replacement therapy ($r = .015$) revealed no significant relationship.

The mean math achievement score for the hypopituitary sample is significantly lower than that of the normative sample (Table 5.3). Twenty-two (52%) earned a standard score <85. The mean reading achievement score is not significantly different from the normative sample; 7 (16%) had standard scores <85. The 7 children with low achievement in reading also had low achievement in math.

Eleven (26%) hypopituitary children have significant visual-motor integration deficits. The Koppitz developmental score for each fell below the 16th percentile and errors included four or more brain injury indicators. Five more children had developmental scores below the 16th percentile but

TABLE 5.3
Comparisons of Hypopituitary and Normative Sample Means for
Cognitive, Affective and Achievement Measures

Measure	Range	GHD Sample $\overline{X}$	SD	Normative $\overline{X}$	SD	T	P
Verbal IQ	52–127	93.9	18.3	100	15	−2.14	<.05
Performance IQ	49–123	94	16.3	100	15	−2.37	<.05
V–P Difference	0– 42	13	9.3	9.7	7.6	2.27	<.05
General Self-Concept	43– 77	60.2	11.4	51.8	13.8	4.59	<.05
Reading	52–131	96.1	16.9	100	15	− .34	NS
Math	55–126	84.7	13.9	100	15	−7.17	<.001

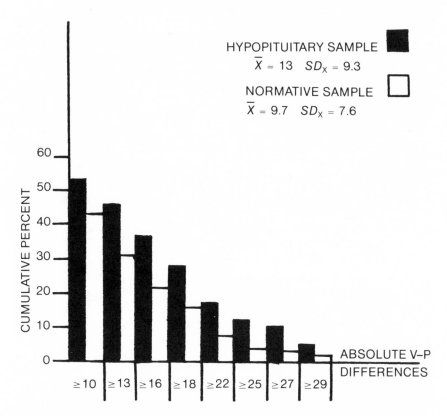

FIG. 5.1.   The incidence of absolute verbal-performance differences for the hypopituitary sample compared to the normative standardization sample.

errors were not suggestive of neurologic immaturity. An additional 16 earned developmental scores between the 16th and 50th percentile with 10 earning scores in the average-high range. These data suggest that a significant number of hypopituitary youngsters have visual-motor integration problems.

The mothers of the children assessed were found to reflect attitudes similar to mothers of normal children. Four mothers supported an attitude consistent with overprotective child-rearing practices, with only one mother endorsing restrictive parenting. These data do not support earlier findings describing mothers of hypopituitary children as infantilizing.

IGHD and MHD children were compared on cognitive, affective, and achievement measures by the Mann Whitney U-Test (Table 5.4). The median V-P difference score of the IGHD group is significantly higher than that of the MHD group. Median VIQ and PIQ scores were not significantly different between groups, but IQs for the MHD group were found to be

TABLE 5.4
Comparisons of Cognitive, Affective, and Achievement Measures for
the IGDH[a] and MHD[b] Children

Measure	IGHD Median	MHD Median	U-Test	P
Verbal IQ	100	86	136.0	NS
Performance IQ	96	87	142.5	NS
V-P Differences	13	7	110.0	<.05
General Self Concept	62	64	151.5	NS
Harsh Punitive Control	4.1	5.1	106.5	<.05
Overpossessiveness	4.5	6.1	103.5	<.05
Reading	98.5	97	171.0	NS
Math	83	78	126.5	NS

[a]IGHD is isolated growth hormone deficiency; [b]MHD is multiple hormone deficiency.

slightly lower. These data suggest that IGHD children have slightly higher but more variable cognitive profiles than MHD children whose profiles are lower and flatter. Mothers of MHD children were significantly more restrictive and overpossessive than mothers of IGHD children. This was not unexpected and may reflect the increased demands placed on the mothers of these more physiologically vulnerable children. The two groups did not differ significantly on self-concept, reading achievement, or median math scores.

Academic achievement among hypopituitary children was of particular interest. Children earning a standard score <85 in either math or reading were identified as low achievers. Twenty-two (52%) children met this criterion with 7 earning low scores in both reading and math, and 15 in math only. Demographically, achievers and low achievers were not significantly different in terms of sex, diagnostic classification, or incidence of medical risk factors. Similarly, group means were not significantly different relative to SES, mother's level of education, chronologic or bone age, age diagnosed, age when treatment was initiated, number of months treated, grade placement, or milestone attainment. Significantly more low achievers than achievers failed a grade.

Low achievers were significantly different from achievers on all cognitive measures but not on affective measures (Table 5.5). Specifically, the mean IQ scale scores, V-P difference scores, and Bender percentiles were significantly lower for the low achievers but mean self-concept and maternal attitude scores were not significantly different. These data suggest that cognitive factors, rather than affective factors, better explain achievement difficulties in hypopituitary children.

The 22 children with low achievement were grouped according to the support for one of the three theories explaining academic failure among

TABLE 5.5
Mean Comparisons of Achievers and Low-Achievers for Cognitive
and Affective Measures

	Achievers (N = 20)		Low-Achievers (N = 22)			
	$\overline{X}$	SD	$\overline{X}$	SD	T	P
Verbal IQ	102.6	11.6	86.7	19.9	3.12	<.01
Performance IQ	100.3	13.1	89.2	17.2	2.33	<.05
V-P Differences	9.6	7.2	15.4	10.4	−2.09	<.05
Bender Percentiles	45.3	29.0	22.9	26.9	2.58	<.01
General Self-Concept	60.5	11.7	59.3	10.9	.34	.73
Harsh Punitive Control	5.18	2.16	4.57	1.74	1.02	.32
Overpossessiveness	5.8	1.96	5.10	1.66	1.26	.21

hypopituitary children. The low ability theory (Table 5.6) was supported by the cognitive profiles of 9 (41%); the cognitive deficit theory (Table 5.7) by 7 (32%) and the cognitive underfunctioning theory (Table 5.8) by 6 (27%). Thus, 16 (73%) of the hypopituitary children with learning problems had at least one significant cognitive atypicality. The six "underfunctioners" did not have restrictive or overpossessive mothers. Their self-concept scores tended to be higher than the group as a whole (x = 68) perhaps reflecting denial or social desirability in responding.

TABLE 5.6
Academic Achievement Commensurate With Low Ability
N = 9 (41%)

Subject #	CA	Diagnostic Classification	VIQ	PIQ	V-P Difference	VMI Deficit
23[b]	8[2]	IGHD	64	81	17	YES
26[c]	8[6]	IGHD	57	78	21	YES
34[c]	9[5]	IGHD	66	49	17	YES
35[c]	10[1]	IGHD	52	74	22	YES
36[a]	9[3]	IGHD	87	85	02	YES
37[a]	13[9]	MHD	86	90	04	NO
38[a]	11[9]	MHD	82	84	02	NO
40[a]	11[8]	MHD	82	76	08	NO
42[b]	14[1]	MHD	67	82	15	YES

Note.   Superscripts indicate months.
[a]Full Scale IQ falls in the Dull Normal Range
[b]Full Scale IQ falls in the Borderline Range
[c]Full Scale IQ falls in the Mildly Impaired Range

TABLE 5.7
Academic Failure Secondary to Cognitive Deficits
N = 7 (32%)

Subject #	CA	Diagnostic Classification	VIQ	PIQ	V-P Difference	VMI Deficit
24	12³	IGHD	109	91	18	NO
27	9¹	IGHD	85	114	29	NO
28	13⁵	IGHD	103	78	25	NO
29	12⁶	IGHD	94	52	42	YES
30	10⁸	IGHD	74	102	28	YES
33	11²	IGHD	73	91	18	YES
39	10⁵	MGHD	82	102	20	NO

Note.   Superscripts indicate months.

TABLE 5.8
Academic Failure Secondary to Cognitive Underfunctioning
N = 6 (27%)

Subject #	CA	Diagnostic Classification	VIQ	PIQ	V-P Difference	VMI Deficit	Self-Concept
21	13⁸	IGHD	101	108	04	NO	69
22	11⁸	IGHD	105	96	09	NO	72
25	13²	IGHD	118	106	12	NO	69
31	13⁸	IGHD	106	106	0	NO	68
32	13⁰	IGHD	127	111	16	NO	62
41	9⁹	MHD	95	101	06	NO	70

Note.   Superscripts indicate months.

DISCUSSION

Our most important finding regarding the cognitive functioning of hypopituitary children is that as a group they have overall average ability, but exhibit specific cognitive atypicalities.Although average intelligence among these children is widely reported (Meyer-Bahlburg et al., 1978; Drash & Money, 1968; Money, 1968), specific cognitive deficits have not been identified. Two atypicalities are identified: significant cognitive variability and visual-motor integration difficulties. Cognitive variability is suggested by the high incidence (29%) of large Verbal-Performance differences on the WISC–R. Significant visual-motor integration difficulty is the second cognitive atypicality identified. Performance on the Bender-Gestalt was significantly low for 11 of the hypopituitary children and low-average for an additional 16. For those children with an identifiable deficit, errors suggestive of neurologic immaturity were also present.

Poor academic achievement is not a result of low self-esteem. The self-concepts for the youngsters in the study were not lower than those for normal children. This finding does not substantiate previous investigations (Kusalic & Fortin, 1975; Rotnem et al., 1977) and may reflect different measurement techniques used or age differences among samples. First, rather than the interviewing and projective techniques used previously, a widely accepted psychometric instrument measuring self-concept was used in this study. Although the former are clinically valuable, especially in assessing individuals, they are less so for research purposes because they yield qualitative data that are difficult to analyze and often subject to examiner bias. Second, the age range in this study was restricted to school-aged children (6-16) whereas most earlier studies reported ages ranging from the preschool years to middle age. It is possible that the more negative psychological consequences of short stature are not manifest until the later teen or young adult years. Studies comparing the self-concepts of hypopituitary subjects of varied age groups are necessary to further clarify this issue.

Poor academic achievement was not found to be related to pathogenic maternal attitudes. Mothers of study children were not found to be more restrictive or overpossessive than mothers of normal children. This finding does not substantiate the results of previous research efforts (Money & Pollitt, 1966; Rotnem et al., 1977) and may again reflect the assessment procedures employed. In past studies, the mothers were assessed exclusively through interviewing. For reasons previously outlined this procedure can be problematic in research. In this study maternal attitudes were measured with a well-known psychometric instrument, allowing comparisons with normative data. However, at least one caveat is recognized: The PARI is known to be influenced by education; less educated mothers reflect more pathogenic attitudes than highly educated mothers (Becker & Krug, 1964). If the mothers who participated in this study were more highly educated than those of hypopituitary children in other samples (maternal education was not reported), the finding that the former are less restrictive and overpossessive may reflect an educational difference.

A high incidence of low academic achievement in hypopituitary children has been reported (Dorner, 1973; Obuchowski et al., 1970; Pollitt & Money, 1964; Spector et al., 1979). The findings of this study substantiate these earlier reports. Twenty-two of the children earned standard scores significantly below average on widely recognized achievement tests in math and/or reading. Previous research efforts have explained school difficulties according to three conflicting theories: the cognitive underfunctioning—low self-concept theory; the low-ability theory, and the cognitive deficit theory. The findings of this study suggest that the low-ability and cognitive deficit theories explain achievement difficulties for 16 children, whereas the cognitive underfunctioning—low self-concept theory best describes poor achievement in 6. These findings have important implications regarding the educational interventions selected for these children. Sixteen of the children in this sample have failed a grade in school. However, psychometric assessment data suggest that retention was not the intervention indicated for 12 of the 16. Specifically, 8 of these children have subaverage (70–89) or impaired (<70) intelligence. The 5 with subaverage ability might have been better served by helping parents and teachers establish more realistic expectations for school performance, and the 3 impaired children required special education placement. An additional 4 of the children retained have cognitive profiles similar to those identified as learning disabled. Remediation, not retention, might have better met their educational needs. Among the 4 remaining, 3 are currently achieving at grade level but 1 continues to do poorly. These findings suggest that the physicians and parents of hypopituitary children with school difficulties should request a comprehensive psychological assessment before retention is considered.

In conclusion, the findings of this study suggest that the psychological profile of hypopituitary children is generally positive: most felt good about themselves and most were achieving at levels commensurate with their ability (20 had no achievement problems; 9 had low ability). Educational remediation may benefit an additional number who appear similar to those described as learning disabled. However, several did function below their cognitive potential. Further research is necessary to better understand the conditions that contribute to cognitive underfunctioning.

ACKNOWLEDGMENTS

We gratefully acknowledge the assistance of Arthur Robin, Children's Hospital of Michigan, in analyzing the data; William Hoffman, Children's Hospital of Michigan, and M. MacGillivary, Children's Hospital in Buffalo, New York, for referral of hypopituitary children for testing; and Beverly Peters and Fae Mallery for typing the manuscript.

REFERENCES

Becker, W.C., Krug, R. S. (1964). The parent attitude research instrument—a research review. *Child Development, 35,* 329–364.

Clements, S. (1966). Minimal brain dysfunction in children. *Monograph of the United States Public Health Service Publication, 3* (1, Serial No. 1415).

Craft, W. H., Underwood, L. E., & Van Wyk, J. J. (1980). High incidence of perinatal insult in children with idiopathic hypopituitarism. *Journal of Pediatrics, 96* (3), 397–402.

Critchley, M. (1970). *The dyslexic child.* Springfield, IL: Thomas.

Dorner, S., & Elton, A. (1973). Short, taught, and vulnerable. *Special Education, 62,* 12–16.

Drash, P. W., & Money, J. (1968). Statural and intellectual growth in congenital heart disease, in growth hormone deficiency, and in sibling controls. In D. B. Cheek (Ed.), *Human Growth. Body composition cell growth, energy, and intelligence* (pp. 606–615). Philadelphia: Lea & Febiger.

Dykman, R., Ackerman, P. T., Clements, S. D., & Peters, J. E. (1971). Specific learning disabilities: An attentional deficit syndrome. In H. Mykelburst (Ed.), *Progress in learning disabilities* (Vol. 2). New York: Grune & Stratton.

Hopwood, N. J., Forsman, P. J., Kenny, F. M., & Drash A. (1975). Hypoglycemia in hypopituitary children. *American Journal of Diseases in Childhood, 129,* 918–926.

Kaevi, A., & Pasamanick, B. (1958). Association of factors of pregnancy with reading disorders in childhood. *Journal of the American Medical Association, 166,* 1420–1423.

Kaufman, A. S. (1975). Factor analysis of the WISC-R at eleven age levels between 6¹/₂ and 16¹/₂ years. *Journal of Consulting and Clinical Psychology, 43,* 135–147.

Kaufman, A. S. (1979). *Intelligent testing with the WISC-R.* New York: Wiley.

Kinsbourne, M. (1973). Minimal brain dysfunction as a neurodevelopmental lag. *Annals of the New York Academy of Sciences, 205,* 268–273.

Koppitz, E. M. (1963). *The Bender Gestalt test for young children.* New York: Grune & Stratton.

Kusalic, M., & Fortin, C. (1975). Growth hormone treatment in hypopituitary dwarfs: Longitudinal psychological effects. *Canadian Psychiatric Association Journal, 20,* 325–331.

Meyer-Bahlburg, H., Feinman, J., MacGillivary, M. H., & Aceto, T. (1978). Growth hormone deficiency, brain development, and intelligence. *American Journal of Diseases of Children, 132,* 565–572.

Michael, W. B., Smith, R. A., & Michael, J. J. (1975). The factorial validity of the Piers-Harris children's self-concept scale for each of three samples of elementary, junior high, and senior high students in a large metropolitan school district. *Educational and Psychological Measurement, 35(2),* 405–414.

Money, J. (1968). Psychological aspects of endocrine and genetic disease in children. In L. I. Gardner (Ed.), *Endocrine and genetic diseases of childhood.* Philadelphia: Saunders.

Money, J., & Pollitt, E. (1966). Studies in the psychology of dwarfism: II. Personality maturation and response to growth hormone treatment in hypopituitary dwarfs. *Journal of Pediatrics, 68,* 381–390.

Mussen, P. H., Conger, J. J., & Kagan, J. (1974). *Child development and personality.* New York: Harper & Row.

Obuchowski, K., Zienkiewicz, H., & Graczykowska-Koczorowska, A. (1970). Psychological studies in pituitary dwarfism. *Polish Medical Journal, 9,* 1229–1235.

Owen, R. P. & Root, A. W. (1979). *Growth hormone deficiency.* Seattle: Human Growth Foundation.

Pollitt, E., & Money, J. (1964). Studies in the psychology of dwarfism. I. Intelligence quotient and school achievement. *Journal of Pediatrics, 64,* 415–421.

Ross, A. (1976). *Psychological aspects of learning disabilities* and reading disorders. New York: McGraw-Hill.

Rotnem, D., Genel, M., Hintz, R. L., & Cohen, D. J. (1977). Personality development in children with growth hormone deficiency. *Journal of the American Academy of Child Psychiatry, 16,* 411–426.

Schaefer, E. S., & Bell, R. Q. (1958). Development of a parental attitude research instrument. *Child Development, 29,* 339–361.

Spector, S., Faigenbaum, D., Sturman, M., & Hoffman, W. (1979, November). *Psychological correlates of short stature in children and adolescents.* Paper presented at the meeting of the American Association of Psychiatric Services for children. Chicago, IL.

6

Long-Term Social Follow-up of Growth Hormone Deficient Adults Treated with Growth Hormone During Childhood

Heather J. Dean

Terri L. McTaggart

David G. Fish

Henry G. Friesen
University of Manitoba, Winnipeg, Manitoba, Canada

The efficacy of human growth hormone (GH) in promoting linear growth in children with GH deficiency (GHD) has been widely confirmed (Frasier, 1983; Guyda, Friesen, Bailey, Leboeuf, & Beck, 1975). Although in the early years of hGH therapy the long-term prognosis for height was unknown, the observed growth acceleration led most investigators, patients, and their families to anticipate normal final adult height (Grew, Stabler, Williams, & Underwood, 1983). Inherent in this prediction was the assumption that improved growth would facilitate normal psychosocial development. Reports of final adult heights of individuals treated with GH during childhood and adolescence are now appearing in the literature (Burns, Tanner, Preece, & Cameron, 1981; Joss, Tuppinger, Schwartz, & Poter, 1983), but the psychosocial status of these patients in adult life has not been studied. Four years ago, in delivering the Raben Memorial Lecture entitled "A Tale of Stature" at the 62nd annual meeting of the Endocrine Society, Henry G. Friesen (1980) noted, "It is unquestionable that growth velocity can be normalized in the majority of children, but it is not altogether clear what the ultimate psychological and social adaptation of these individuals is." To pursue this question, we interviewed 116 growth hormone deficient (GHD) adults across Canada (aged 18–40 years) who had been treated for at least 2 years with growth hormone during childhood. Our objective was to determine their social status using education, employment, and marital status as outcome variables. The null hypothesis of our study was that the educational, employment, and marital status of the GHD adults was not different from that of the general population. Regional population data from Statistics Canada (1980, 1983) were available for comparison. Parents and siblings

73

over 18 were used as environmental control subjects. This presentation is a summary of the results of that study (Dean, McTaggart, Fish, & Friesen, 1984).

STUDY DESIGN

Since 1967 in Canada, evaluation of all applications for growth hormone therapy, distribution of growth hormone, biannual collection of clinical data and analysis of these data have been performed in a central facility (Guyda et al., 1975). Using this information, we were able to assess the influence of 12 treatment variables on social status (Table 6.1). Of the 179 subjects eligible for the study, we were able to contact 116 with all agreeing to participate. The interviewed group was not different from the non-interviewed group in terms of the 12 treatment variables.

Of the 116 patients interviewed, 86 were males and 30 females. Sixty-eight percent had idiopathic isolated GHD; 80% had a non-organic cause for their hypopituitarism (Table 6.2). The average age at initiation of treatment was 13.6 years (Table 6.3); however, there was a wide spread with the youngest patient at initiation being 4 years and the oldest 23. The mean duration of treatment was 6.0 years, with a range of 2 to 15 years. The average number of standard deviations below the normal mean height for age and sex before therapy (i.e., standard deviation score or SDS-ht.) was 4.4, with a range 1.7–9.2. The mean SDS-ht. had increased to 3.0 after therapy and the average height change afforded by GH therapy was an increase of 1.5 standard deviations toward the mean height.

RESULTS

Ninety-six subjects had finished their formal education; 20 were still students (Table 6.4). Thirty-six had not completed high school; 52 had finished high school and, of these, 20 had a post-secondary school diploma. One of the 20 post-secondary diplomas was a university degree. This distribution is similar to the Canadian population except that the ratio of vocational to university post-secondary school diplomas is higher for our patient group. The distribution of the educational level achieved was similar for the idiopathic GHD and organic GHD. In general, the educational achievement of the patients was higher than their parents and similar to their siblings. Educational achievement was not affected by any of the 12 treatment variables.

The rate of employment of the GH graduates was less than expected (Fig. 6.1), with 35% being unemployed. Those individuals less than 25 years

TABLE 6.1
Treatment Variables

1. Age at initiation of therapy
2. Age at termination of therapy
3. Duration of therapy
4. Mean parental height
5. Standard deviation score for height (SDS-ht.)[a] prior to therapy
6. SDS-ht. after termination of therapy
7. Δ SDS-ht.
8. Other hormone replacement
9. Other physical abnormalities
10. Primary diagnosis
11. Treatment for organic lesions
12. Reason for termination of therapy

[a]The SDS-ht. was calculated using the formula:

$$\frac{\text{(mean height for age and sex) minus (actual height)}}{SD \text{ of mean height for age and sex}}$$

TABLE 6.2
Primary Diagnosis of 116 GHD Subjects

Diagnosis	No. of Patients	
Organic	25	
Craniopharyngioma		16
Post Irradiation		4
Post Traumatic		3
Chromophobe Adenoma		1
Histiocytosis-X		1
Idiopathic	91	
Isolated GHD		73
Hypopituitarism		12
Familial		6

of age were most severely affected, even after controlling for the expected age-related bias toward the young in the general population. None of the other treatment variables, nor sex, place of residence or education, affected employment status. There appeared to be a trend toward higher unemployment with increasing SDS-ht. (i.e., shorter stature) after therapy in those less than 25 years of age, but this was not statistically significant. The unemployment rate was higher in the patients than their parents and siblings.

The majority (85%) of our patients have never been married (Fig. 6.2). Although the mean age of our group was 25, this outcome could not be explained on the basis of age because the marriage rate at each 5-year age

TABLE 6.3
Demographic Data and Final Results of 116 GHD Patients

	Range of All Subjects (N = 116)	Mean ± SD		
		All Subjects (N = 116)	Males (N = 86)	Females (N = 30)
Age at Commencement GH (yrs.)	4.0–22.8	13.6 ± 4.0	13.9 ± 3.8	12.9 ± 4.4
Age at Termination GH (yrs.)	12.3–26.4	18.7 ± 2.9	18.9 ± 3.0	18.0 ± 2.6
Number of Treatment Courses	2–15	6.0 ± 2.9	6.1 ± 3.0	6.0 ± 2.8
Pretreatment SDS-height	1.7–9.2	4.4 ± 1.7	4.3 ± 1.7	4.8 ± 1.8
Posttreatment SDS-height	0.7–7.2	3.0 ± 1.2	3.1 ± 1.2	2.8 ± 1.3
Change in SDS-height	−1.5–5.8	1.4 ± 1.3	1.2 ± 1.3	2.0 ± 1.4
Midparental stature (cm)	147.4–179.1	165.5 ± 5.5	165.0 ± 5.7	167.1 ± 4.8

TABLE 6.4
Education Achievements of 116 GHD Subjects

		Diagnosis		
Diploma	Total Group N (%)	Idiopathic N (%)	Cranio- Pharyngioma N (%)	Other N
None	36 (38)	25 (33)	7 (50)	4
Secondary School (S.S.)	32 (33)	30 (39)	2 (14)	0
Vocational				
with S.S.	19 (20)	13 (17)	4 (29)	2
without S.S.	8 (8)	7 (10)	1 (7)	0
University	1 (1)	1 (1)	0	0
Subtotal	96	76	14	6
Student S.S.	3	—	1	2
Post S.S.	17	15	1	1
Total	116	91	16	9

interval was significantly lower than that observed in the general population. None of the 12 treatment variables affected marital status. Although there was a trend for more of the hypogonadotropic individuals who were 25 years or older to be single, this result did not reach statistical significance (Table 6.5).

Seventy of the 96 subjects who had completed their formal schooling

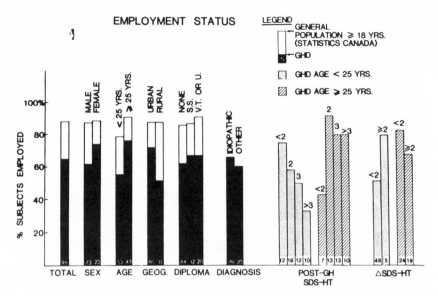

FIG. 6.1.   The employment status of the 96 GHD adults in the work force
expressed as percentage employed (solid bars). Comparison of employment
status to the general population (open bars) is shown where possible. Of the
treatment variables studied, only diagnosis, post-treatment SDS-height, and
the increase in SDS-height are illustrated. The only statistically significant
relationship is with age ($p < .05$). The number of subjects in each group is
indicated at the bottom of each bar. Geog, geography; SS, secondary school;
VT, vocational training; U, university; SDS-ht, SDS-height.

lived at home (Table 6.6). Excluding the 6 patients with decreased vision,
only 58% had a driver's license. Twenty-four had seen a psychologist or
psychiatrist for stature-related problems during the time that they received
GH therapy, and 27 of the parents of the GHD subjects were separated or
divorced. However, these two subgroups were not different from the rest of
the GHD subjects in terms of employment or marital status. The majority
(85%) of respondents subjectively felt that height was not a serious problem
to them at present.

DISCUSSION

The management of GHD in childhood has focused almost exclusively
on increasing the height of the affected child. The results of this study
indicate that although increased final adult height, measured as an absolute
decrease in SDS-ht., has been achieved, the overall outcome of therapy is
unsatisfactory. Similar results have been found in three unpublished surveys (one from France [F. Bouchayer, personal communication, August,

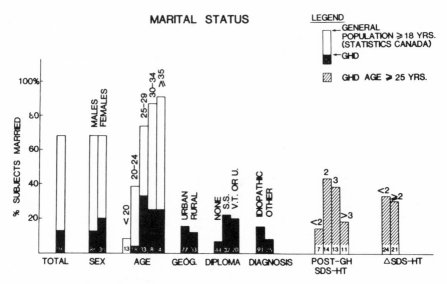

FIG. 6.2.   The marital status of the 116 GHD adults expressed as percentage married or divorced (solid bars). Comparison of the marital status to the general population (open bars) is shown where possible. The number of subjects in each group is indicated at the bottom of each bar. Note that only the 96 patients who had completed their education have a final diploma and that only the 45 patients older than 25 years have been included for post-GH SDS-height or SDS-height. The only statistically significant relationship is with age ($p < .01$). Geog, geography; SS, secondary school; VT, vocational training; U, university; SDS-ht, SDS-height.

TABLE 6.5
Marital Status of 116 GHD Subjects with
Respect to Gonadal Function and Age

Age	Gonadal Function	Total No. of Subjects	Married or Divorced N (%)	Single N (%)
< 25 years	Hypogonadotropism	22	0	22 (100)
	Normal	49	2 (4)	47 (96)
≥ 25 years[a]	Hypogonadotropism	21	4 (19)	17 (81)
	Normal	24	10 (42)	14 (58)

[a]Fisher's Exact Test (one-tailed), $p = .09$

TABLE 6.6
Miscellaneous Social Parameters

Living At Home With Parents	70/ 96	(73%)
Driver's License	67/116	(58%)
Psychological Counseling	24/116	(21%)
Family Discord	26/117	(23%)

1984] and two from the United States reported in this volume by R. Clopper and A. J. Johanson). Joss et al. (1983) reported the final adult heights of 18 Swiss GHD patients and commented that none of these individuals, all older than 20 years, were married. We do not know if these results are unique to GHD individuals, if they simply reflect the impact of a chronic disease on psychosocial development during childhood, adolescence or adulthood or if there are undefined factors involved (Table 6.7).

TABLE 6.7
Possible Factors Related to Social Outcome of GHD Adults

1. Chronic Disease
2. Short Stature
3. Visibility
4. Adolescence

A fundamental question that must be addressed is whether GHD can be considered a chronic disease. By definition, a chronic disease is not curable, at least in the short term. In general, GHD has been considered a chronic problem rather than a disease, insofar as long-term therapy and medical follow-up are required. It has not, however, been in the same category as diabetes, asthma, and cystic fibrosis as it is not life-threatening and is potentially curable. However, curability is uncertain because the final adult heights observed to date are disappointing, with the mean final heights of treated patients being 2-3 standard deviations below the normal mean. Thus, if GHD is included with all of the chronic diseases of childhood, we can begin to explore some of the factors causing social maladjustment in adult life. The psychosocial issues surrounding such chronic childhood diseases as diabetes, arthritis, cancer, Crohn's disease or ulcerative colitis, and cystic fibrosis have been studied in isolation for years. The concept that each of these diseases has a specific effect on personality development, family dynamics, and ultimate social integration has been challenged during the past decade. A unifying concept has emerged that the spectrum of personality traits is similar for *all* chronic diseases, and that the burdens and stresses of each of these diseases ultimately have the same influence on developmental processes (Coupey & Cohen, 1984). It has been estimated that the risk of significant psychological or social problems in children with chronic physical disorders is 1.5–3 times higher than the normal population (Pless, 1984). Nevertheless, the rates of employment and marriage for adults with various chronic diseases since childhood are higher than our GHD population (Heller, Tidmarsh, & Pless, 1981; Hill, Herstein, & Walters, 1976; Marks, Barnett, & Calis, 1982; McWilliams & Paradise, 1973; Miller, Spitz, Simpson, & Williams, 1982). This discrepancy may be related to the personality characteristics ascribed to children with GHD; namely, that they tend to be withdrawn, socially isolated, and powerless, to lack normal aggression, and to have low self-esteem (Krims, 1968; Kusalic & Fortin, 1975; Kusalic, Fortin, & Gauthier, 1972; Money & Pollitt, 1966; Rotnem, Cohen, Hintz, & Genel, 1979; Rotnem, Genel, Hintz, & Cohen, 1977). These findings must be contrasted with the normal results of psychological testing reported by C. S. Holmes and P. T. Siegal in this volume and by others (Abbott, Rotnem, Genel, & Cohen, 1982; Drotar, Owens, & Gotthold, 1980; Stabler & Underwood, 1975).

The issue of visibility of disease has not been specifically studied. It is possible that visibility of a chronic disease may be inversely related to rates of employment and marriage. In support of this hypothesis, Cassileth et al. (1984) found that the rate of marriage for adults with dermatologic disorders was less than the rate of marriage for groups of adults with other chronic diseases such as arthritis, diabetes, cancer and renal disease. In GHD the visibility factor may be magnified by age because most of our patients were diagnosed and treated during the vulnerable adolescent years. At this age, short stature and delayed pubertal development may have profound effects on three of the most critical developmental processes, namely, definition of one's identity, sexuality, and independence (Coupey & Cohen, 1984). It is noteworthy that 75% of our adult respondents complained that people still mistook them for younger than their age even though their final height was satisfactory.

In summary, our hypothesis that GHD treated individuals would be socially integrated has been disproven. Although educational achievements are similar to the general population and siblings, the rates of employment and marriage are significantly lower than the general population and siblings. It remains to be determined whether this apparent social maladjustment is similar to children with other chronic medical disorders or whether there are factors unique to GHD, such as high visibility and delayed puberty, which make them at higher risk. As we found no correlation between height and employment or marriage, it may be simplistic to assume that earlier diagnosis and treatment, the use of different GH treatment schedules, and the achievement of greater final adult height will have any positive impact on social outcome. The impact of earlier induction of puberty in gonadotropin deficient patients remains to be determined.

In conclusion, the high rate of unemployment and low rate of marriage of GHD adults is sobering and should be considered seriously in any future deliberations regarding the use of GH for healthy short children (Gertner et al., 1984; Van Vliet, Styne, Kaplan, & Grumbach, 1983). The media are already claiming physical and social benefits of GH in these children (Chase, 1983). GH therapy could focus undue attention on height and create unrealistic expectations in these children (Grew et al., 1983). These caveats are particularly timely in view of the potential for future indiscriminate use of recombinant GH.

ACKNOWLEDGMENTS

This research was supported by grants from the Manitoba Medical Services Foundation and the Medical Research Council of Canada (MA-2525) and a Research Fellowship of the Medical Research Council of Canada (H.J. Dean).

REFERENCES

Abbott, D., Rotnem, D., Genel, M., & Cohen, D. J. (1982). Cognitive and emotional functioning in hypopituitary short-statured children. *Schizophrenia Bulletin, 8,* 310-319.

Burns, E. C., Tanner, J. M., Preece, M. A., & Cameron, N. (1981). Final height and pubertal development in 55 children with idiopathic growth hormone deficiency treated for between 2 and 15 years with human growth hormone. *European Journal of Pediatrics, 137,* 155-164.

Cassileth, B. R., Lusk, E. J., Strouse, T. B., Miller, D. S., Brown, L. L., Cross, P. A., & Tenaglia, A. N. (1984). Psychosocial status in chronic illness. A comparative analysis of six diagnostic groups. *New England Journal of Medicine, 311,* 506-511.

Chase, M. (1983, June 24). *Wall Street Journal,* New York, p. 1.

Coupey, S. M., & Cohen, M. I. (1984). Special considerations for the health care of adolescents with chronic illnesses. *Pediatric Clinics of North America, 31,* 211-219.

Dean, H. J., McTaggart, T. L., Fish, D. G., & Friesen, H. G. (1984). *Evaluation of the educational, vocational and marital status of growth hormone deficient (GHD) adults treated with growth hormone during childhood* (submitted for publication).

Drotar, D., Owens, R., & Gotthold J. (1980). Personality adjustment of children and adolescents with hypopituitarism. *Child Psychiatry and Human Development, 11,* 59-66.

Frasier, S. D. (1983). Human pituitary growth hormone (hGH) therapy in growth hormone deficiency. *Endocrine Reviews, 4,* 155-170.

Friesen, H. G. (1980). A tale of stature. *Endocrine Reviews, 1,* 303-318.

Gertner, J. M., Genel, M., Gianfredi, S. P., Hintz, R. L., Rosenfeld, R. G., Tamborlane, W. V., & Wilson, D. M. (1984). Prospective clinical trial of human growth hormone in short children without growth hormone deficiency. *Journal of Pediatrics, 104,* 172-176.

Grew, R. S., Stabler, B., Williams, R. W., & Underwood, L. E. (1983). Facilitating patient understanding in the treatment of growth delay. *Clinical Pediatrics, 22,* 685-690.

Guyda, H. J., Friesen, H. G., Bailey, J. D., Leboeuf, G., & Beck, R. M. (1975). Medical Research Council of Canada therapeutic trial of human growth hormone: First 5 years of therapy. *Canadian Medical Association Journal, 112,* 1301-1309.

Heller, A., Tidmarsh, W., & Pless, I. B. (1981). The psychosocial functioning of young adults born with cleft lip or palate. *Clinical Pediatrics, 20,* 459–465.

Hill, R. H., Herstein, A., & Walters, K. (1976). Juvenile rheumatoid arthritis—follow-up into adulthood. *Canadian Medical Association Journal, 114,* 790–796.

Joss, E., Tuppinger, K., Schwartz, H. P., & Poter, H. (1983). Final height of patients with pituitary growth failure and changes in growth variables after long term hormonal therapy. *Pediatric Research, 17,* 676–679.

Krims, M. B. (1968). Observation on children who suffer from dwarfism. *Psychiatric Quarterly, 42,* 430–443.

Kusalic, M., & Fortin, C. (1975). Growth hormone treatment in hypopituitary dwarfs: Longitudinal psychological effects. *Canadian Psychiatric Association Journal, 20,* 325–331.

Kusalic, M., Fortin, C., & Gauthier, Y. (1972). Psychodynamic aspects of dwarfism. Response to growth hormone treatment. *Canadian Psychiatric Association Journal, 17,* 29–34.

Marks, S. H., Barnett, M., & Calis, A. (1982). A case-controlled study of juvenile and adult onset ankylosing spondylitis. *Journal of Rheumatology, 9,* 739–741.

McWilliams, B. J., & Paradise, L. P. (1973). Educational, occupational and marital status of cleft palate adults. *Cleft Palate Journal, 10,* 223–228.

Miller, J. J., Spitz, P. W., Simpson, U., & Williams, G. F. (1982). The social function of young adults who had arthritis in childhood. *Journal of Pediatrics,* 378–382.

Money, J., & Pollitt, E. (1966). Studies in the psychology of dwarfism. II. Personality maturation and response to growth hormone treatment in hypopituitary dwarfs. *Journal of Pediatrics, 68,* 381–390.

Pless, I. B. (1984). Clinical assessment: Physical and psychological functioning in chronic disease in childhood. *Pediatric Clinics of North America, 31,* 33–45.

Rotnem, D., Cohen, D. J., Hintz, R., & Genel, M. (1979). Psychological sequelae of relative "treatment failure" for children receiving human growth hormone replacement. *Journal of the American Academy of Child Psychiatry, 18,* 505–520.

Rotnem, D., Genel, M., Hintz, R., & Cohen, D. J. (1977). Personality development in children with growth hormone deficiency. *Journal of the American Academy of Child Psychiatry, 16,* 412–426.

Stabler, B., & Underwood, L. E., (1977). Anxiety and locus of control in hypopituitary dwarf children. *Research Relating to Children Bulletin, 38,* 75.

Statistics Canada. (1980). Standards Division. *Social concepts directory: A guide towards standardization in statistical surveys.*

Statistics Canada. (1983). *Labour force survey* (Catalogue 71-001, Vol. 39, No. 5).

Van Vliet, G., Styne, D. M., Kaplan, S. L., & Grumbach, M. M. (1983). Growth hormone treatment for short stature. *New England Journal of Medicine, 309,* 1016–1022.

7 Post-treatment Follow-up of Growth Hormone Deficient Patients: Psychosocial Status

Richard R. Clopper

Margaret H. MacGillivray

Tom Mazur

Mary L. Voorhess

Barbara J. Mills
*State University of New York at Buffalo
and Children's Hospital of Buffalo, New York*

Human growth hormone has been used for over 25 years in the treatment of growth hormone (GH) deficiency. During this time significant knowledge has been gained about the somatic effects of GH in individuals with short stature. In addition, data on the IQ and cognitive abilities (Abbott, Rotnem, Genel, & Cohen, 1982; Clopper et al., 1977; Meyer-Bahlburg, Feinman, MacGillivray, & Aceto, 1978; Pollitt & Money, 1964; Rosenbloom, Smith, & Loeb, 1966), academic achievement (Holmes, Hayford, & Thompson, 1982; Pollitt & Money, 1964; Rosenbloom et al., 1966), personality characteristics (Drotar, Owens, & Gotthold, 1980; Rotnem, Genel, Hintz, & Cohen, 1977), psychosocial function (Dorner & Elton, 1973; Ehrhardt & Meyer-Bahlburg, 1975; Meyer-Bahlburg, 1985; Weiss, 1977), and psychosexual status (Clopper, Adelson, & Money, 1976; Meyer-Bahlburg & Aceto, 1976; Money, Clopper, & Menefee, 1980) of GH deficient individuals have accumulated on patients during or just completing their GH treatment. Data on their long-term, post-treatment functioning are just now becoming available. This report presents the first behavioral data from a recent follow-up study of the original cohort of hypopituitary patients who were treated with human growth hormone (GH)[1] at the Children's Hospital of Buffalo. Our question was, "How are these individuals functioning now after completing GH therapy?" Past studies of the behavioral effects of GH treatment have been hindered by small sample size, the clinical heterogeneity of the study groups, and the lack of uniform behavioral measures (Mazur & Clopper, in press). In addition to the large number of participants, this study differs from previous ones in that the dose of GH was standardized throughout for each patient, and the study groups were more homogeneous with respect to their endocrine status.

DESCRIPTION OF SAMPLE

The sample consists of 39 patients (7 female, 32 male) who completed their treatment with GH prior to December 31, 1982.The group ranged in age from 17 to 34 years at follow-up, with a mean age of 23 ± 4.5 years. All 39 gave written informed consent to participate in this study. Of the 39 respondents, 18 were diagnosed as having idiopathic isolated GH deficiency (IGHD) (3 female, 15 male), and 21 patients as having multiple pituitary deficits (MPD) (4 female, 17 male). The frequency of these pituitary deficits and additional demographic data are given in Table 7.1. Almost half of the group with multiple pituitary deficits had tumor-related hypopituitarism (10 cases). Six patients (33%) with isolated growth hormone deficiency were considered partially growth hormone deficient. Growth hormone treatment consisted of 0.1 u/kg I.M. 3 times per week for 9 of 12 months until the patient reached the maximum height permitted by NHPP criteria. This maximum height criterion changed periodically. Thus patients who received HGH therapy recently had longer treatment periods than those who received it during the early 1970s.

The mean chronologic age (CA) and bone age (BA) at onset of GH treatment were similar for the IGHD and MPD groups (Table 7.1). The mean chronologic age at the end of GH treatment was also similar for both groups (CA 17.1 vs. 18.1 years). However, at the end of the treatment period, the MPD group had a younger mean bone age than the IGHD group (Table 7.1). The mean age at the time of follow-up was significantly greater in the MPD group (25.5 ± 4.6 years) as compared to the IGHD group (20.9 ± 2.9 years). The mean duration of treatment in the IGHD patients was 5.6 years compared to 7 years in the MPD patients. This difference is not statistically significant.

Pre- and post-treatment heights were similar in the two diagnostic groups (Table 7.1). Mean heights of the two groups at the time of the survey were

---

[1]Supplied by the National Hormone and Pituitary Program (NHPP), formerly the National Pituitary Agency.

TABLE 7.1
Demographic Data

Patient Groups	Mean Age (years)			Mean Duration of therapy (yr)	Mean Height (SD Units)[d]			Mean Weight (SD Units)[d]			Number of Cases with pituitary deficits				
	Start GH Treatment (BA)[a]	End (BA)[a]	Question-naire		Start GH Treatment	End	Question-naire	Start GH Treatment	End	Question-naire	GH	TSH	ACTH	Gn	ADH
Isolated GH Deficiency (IGHD) [n = 18]	11.4 (8.3)	17.1 (15.2)[c]	20.9[b]	5.6	-3.9	-2.4	-1.9	-1.8	-0.9	-0.5	18	0	0	0	0
Multiple Pituitary Deficits (MPD) [n = 21]	12.6 (7.9)	18.1 (13.6)[c]	25.5[b]	7.0	-4.1	-2.0	-1.2	-1.8	-1.0	-0.08	21	17	16	19	5
Total Sample [n = 39]	12.1 (8.1)	17.6 (14.3)	23.4	5.6	-4.0	-2.2	-1.5	-1.8	-0.9	-0.3	39	17	16	19	5

[a]BA = Bone Age

[b]t = -3.6, with 37 degrees of freedom, p = .001

[c]t = 2.9 with 32 degrees of freedom, p = .006

[d]the norms for height and weight are from Tanner, J. M., et al. (1966). *Archives Diseases of Childhood, 41*, 454–471 and 613–635.

also not different. Females in the IGHD group reported a mean adult height of 58.2 in. at follow-up, whereas their male counterparts reported a mean height of 63.9 in. The females in the MPD group reported a mean height of 61.2 in. but the MPD males reached a mean height of 65.5 in. The specific deficits of the patients with MPD are given in Table 7.1.

METHOD AND PROCEDURES

Each of the 39 participants in this study completed a 39-item questionnaire derived from the structured interview schedule used by Money et al.(1980) and Clopper, Mazur, MacGillivray, Peterson, and Voorhess (1983). The instrument required patients to recall facts of their medical history and behavioral development. In addition they were asked to estimate their current frequency of various work, recreational, and social activities. In addition to the questionnaire data, each patient's medical history was abstracted. The resulting data were computerized for subsequent analysis using the *Statistical Package for the Social Sciences* (Nie, Hull, Jenkins, Steinbrenner, & Bent, 1975).

A total of 80 patients were identified as having completed their treatment with GH through the Children's Hospital of Buffalo. Of those 80, 13 patients could not be located, leaving 67 subjects who were contacted by mail and/or phone and asked to participate in the study. Fifty-six patients gave written informed consent and completed the questionnaire for a response rate of 84% of those contacted. Of the 56 participants, 17 patients were excluded from the analysis because they either had normal growth after GH was discontinued, or they were treated for less than 6 months. The 39 patients considered in this report represent 78% of the 50 contacted patients who were proven to be GH deficient by standard diagnostic criteria and who had completed their entire course of GH treatment.

RESULTS

Educational Experience

Tables 7.2 and 7.3, respectively, give data on highest educational degree attained and the number of subjects who repeated grades during their basic education (kindergarten through grade 12). Ninety-five percent ($n = 37$) of the total sample completed high school and 70% ($n = 26$) of the high school graduates continued their formal education or training beyond the high school level. For 62% ($n = 23$) of the high school graduates their advanced training included college matriculation. Of the two groups significantly more of the older MPD patients attained college experience and degrees than the IGHD patients.

The data on grades repeated during the years of basic education are given in Table 7.3. Although missing data restrict generalization, an interesting trend emerges. Sixteen patients (41% of the total sample) repeated at least one grade; 4 patients (10%) repeated more than one grade. No difference was noted between the subgroups in the frequency of the grade failures. The elementary school grades, however, clearly predominated as the most frequently repeated grades. Thirteen of the 16 patients (81%) repeating any grade repeated a primary school grade. Furthermore, of the 13 patients who repeated an elementary school grade, 8 (62%) repeated either kindergarten or first grade. A similar experience was reported by the 4 patients (3 MPD, 1 IGHD) who repeated more than one grade during their kindergarten through grade 12 education. In 3 of these 4 patients, the initial failure was kindergarten or first grade.

Current Employment and Work

Of the subjects, 54% ($n = 21$) were employed full-time in either service or professional jobs.(Table 7.4) Service positions included work as store clerks and sales people, restaurant cook, construction and skilled labor, members of the armed forces, teacher or rehabilitation aid, and housewife. Professional positions were as computer operator, teacher, musician, hospital administrator, manager of retail store, design engineer, and dairy farmer. An additional 15 patients (39%) were students and not yet employed full-time. Only 8% ($n = 3$) of the total sample were unemployed and not pursuing additional education or training. This rate of unemployment occurred at a time when unemployment figures for western New York were in the 9% to 10% range. The differences between the two subgroups on employment category are not significant.

The median income figures in Table 7.4 reflect the reports of only 18 subjects (46% of total) and show no significant differences between the two subgroups. The range for those reporting their gross annual income was $3,100–$100,000. In western New York during the survey period, the median annual income for households was $17,119 and for unrelated individuals older than 15 was $6,050 (New York State Department of Commerce, 1982). When asked if they liked their present job, 89% ($n = 24$) of the 27 who responded reported that they did like their work.

TABLE 7.2
Highest Degree Completed

Patient Groups	Did not complete High School	Completed High School	Received Technical Training	Some College Experience	Associate's Degree	Bachelor's Degree	Master's Degree
Isolated GH Deficiency[a] (IGHD) [n = 18]	2 (11%)	7 (40%)	3 (17%)	4 (22%)	None	2 (11%)	None
Multiple Pituitary Deficits[a] (MPD) [n = 21]	None	4 (19%)	None	6 (29%)	7 (33%)	2 (10%)	2 (10%)
Total Sample [n = 39]	2 (5%)	11 (28%)	3 (8%)	10 (26%)	7 (18%)	4 (10%)	2 (5%)

[a] X^2 = 15.1, with 7 degrees of freedom, p = .03

TABLE 7.3
Individuals Reporting Repeated Grades (K–12 Grades)

Patient Groups	Subjects repeating any grade (% of total)	Individuals reporting repeated grades K-6 7-12 (% of respondents)	
Isolated GH Deficiency	8 (44%)	6 (75%)	2 (25%)[a]
(IGHD)	(nr = 8)	(nr = 8)	
[n = 18]			
Multiple Pituitary Deficits	8 (38%)	7 (88%)	3 (38%)[a]
(MPD)	(nr = 8)	(nr = 8)	
[n = 21]			
Total Sample	16 (41%)	13 (81%)	5 (31%)[a]
[n = 39]	(nr = 16)	(nr = 16)	

Note.   Figures are for total sample except where indicated by nr = number reporting.
[a]Figures do not total to nr; 4 subjects repeated more than one grade.

TABLE 7.4
Current Employment Status and Income Data

Patient Groups	Gross Yearly Median Income	Unemployed	Employment Student	Category Service Industry	Professional
Isolated GH Deficiency (IGHD) [n = 18]	$10,800 (nr = 7)	1 (6%)	8 (44%)	7 (39%)	2 (11%)
Multiple Pituitary Deficits (MPD) [n = 21]	$12,000 (nr = 11)	2 (10%)	7 (33%)	6 (29%)	6 (29%)
Total Sample [n = 39]	$11,093 (nr = 18)	3 (8%)	15 (39%)	13 (33%)	8 (21%)

Note.   Figures are for total sample except where indicated by nr = number reporting.

Current Living Situation

Table 7.5 indicates that 13 patients (33%) were living independently of their parents at the time of the survey. These 13 patients were equally split between living (1) alone; (2) with a same-sex roommate; or (3) with a spouse. Four of the total sample (10%) were married (3 IGHD and 1 MPD). The mean age at time of marriage was 22.8 ± 2.2 years. Although there was no difference between the subgroups in terms of the number of individuals living on their own, the MPD subjects who had left home did so at a significantly older age than their IGHD counterparts (Table 7.5).

Free Time Social Activities

Table 7.6 shows that the entire sample reported spending an average of 55% of their free time in activities with at least one other person and that the two subgroups did not differ significantly in this respect. Sixty-seven percent of the total ($n = 26$) reported having between 6 and 10 close friends. The majority of the sample ($n = 34$, 87%) had dating experience. Only 8 (38%) of the 21 subjects who responded had a current sexual partner and these figures include the 4 married subjects. Although the differences between the subsamples with respect to current sexual partner do not quite reach statistical significance, the trend is clearly toward more current sexual experience among the IGHD subgroup.

Juvenilization by Others

Because juvenilization has been a recurrent theme in the behavioral literature, our subjects were asked if they experienced being treated younger than their chronologic age in the past and at present.Half the total sample ($n = 19$) reported that they presently experience juvenilization. (Table 7.7) Based on their past experience, juvenilization was most likely to occur by strangers, followed by peers or friends and least likely to occur by family members. Nonetheless, family members do not escape the propensity to juvenilize a GH deficient patient, because 25% ($n = 9$) of those responding

TABLE 7.5
Current Living Situation

Patient Groups	Number (%) Living independently	Age when moved out on own (years)	Number (%) Married
Isolated GH Deficiency (IGHD) [n = 18]	6 (33%)	18.7 ± 1.8[a] (nr = 6)	3 (17%)
Multiple Pituitary Deficits (MPD) [n = 21]	7 (33%)	22.3 ± 3.0[a] (nr = 7)	1 (5%)
Total Sample [n = 39]	13 (33%)	20.2 ± 3.3 (nr = 13)	4 (10%)

Note.   Figures are for total sample except where indicated by *nr* = number reporting.
[a]$t = -3.0$, with 12 degrees freedom, $p = .011$

TABLE 7.6
Free Time Social Activities

Patient Groups	Mean ± SD % Free time spent with others	Number (%) with Dating Experience	Number (%) with sexual partner at follow-up
Isolated GH Deficiency (IGHD) [n = 18]	60.5 ± 24.1	17 (94%)	7 (54%) (nr = 13)
Multiple Pituitary Deficits (MPD) [n = 21]	51.0 ± 22.8	17 (81%)	1 (12%) (nr = 8)
Total Sample [n = 39]	55.4 ± 23.6	34 (87%)	8 (38%) (nr = 21)

Note.   Figures are for total sample except where indicated by nr = number reporting.

TABLE 7.7
Patient Reported Juvenilization

Patient Groups	Juvenilized Now (number; %)	Past Juvenilization		
		by family (number; %)	by friends (number; %)	by strangers (number; %)
Isolated GH Deficiency (IGHD) [n = 18]	7 (41%) (nr = 17)	4 (23%) (nr = 17)	5 (28%)	10 (62%) (nr = 16)
Multiple Pituitary Deficits (MPD) [n = 21]	12 (57%)	5 (26%) (nr = 19)	9 (47%) (nr = 19)	14 (70%) (nr = 20)
Total Sample [n = 39]	19 (50%) (nr = 38)	9 (25%) (nr = 36)	14 (38%) (nr = 37)	24 (67%) (nr = 36)

Note.   Figures are for total sample except where indicated by nr = number reporting.

to the question experienced at least one incident of juvenilization by a family member.

Self-Perception of Physical Appearance

In an effort to assess patients' satisfaction with their physical appearance, patients were asked to estimate how old they believe they look now and to list any dissatisfactions they might have with their appearance. Table 7.8 provides a summary of these data. The first column gives the average magnitude (years) of discrepancy between patients' chronologic age (CA) and their self-estimated age of appearance (SAA). The discrepancy score was obtained by subtracting the SAA from the CA for each patient. The total sample reported that they looked younger than their CA by an average of 5 years. This figure masks the significant difference between the discrepancy perceived by the sample subgroups. The MPD patients perceived a larger discrepancy (6.5 years) between their SAA and CA than did the IGHD patients (3.0 years).

Approximately 44% of the respondents (n = 27) were satisfied with their appearance. Of those who stated some dissatisfaction with their appearance (n = 12), the most frequently cited problem for the females was weight and the least cited was height. For the males, dissatisfaction with their young appearance, degree of virilization, and suboptimal muscle tone were the most frequently cited.

Counseling Experience

Counseling experience was assessed in terms of the number of patients with counseling experience (Table 7.9), and the concerns about which patients considered counseling, regardless of whether or not they actually entered counseling (Table 7.10). Approximately 38% (n = 15) of the total sample reported experience with counseling prior to follow-up. The subgroups did not differ significantly in this respect. The data on current counseling experience are more difficult to interpret due to missing observations. As reported, however, less than 30% of the sample are currently in counseling and the experience of the two subgroups is not significantly different (Fisher Exact Probability, p = .18).

TABLE 7.8
Patients' Perception of Their Appearance

Patient Groups	Mean discrepancy between self-estimated age of appearance and CA[a]	Satisfied with appearance (number %)
Isolated GH Deficiency (IGHD) [n = 18]	3.0[b] years (nr = 17)	7 (47%) (nr = 15)
Multiple Pituitary Deficits (MPD) [n = 21]	6.5[b] years	5 (42%) (nr = 12)
Total Sample [n = 39]	5.0 years (nr = 38)	12 (44%) (nr = 27)

Note.   Figures are for total sample except where indicated by nr = number reporting.
[a]Discrepancy Score = CA − SSA
[b]t = −2.98, with 36 degrees freedom, p = .005

TABLE 7.9
Counseling Experience

| | Cases Reporting | |
Patient Groups	Any Counseling Experience	Current Counseling
Isolated GH Deficiency (IGHD) [n = 18]	4 (22%)	None (nr = 6)
Multiple Pituitary Deficits (MPD) [n = 21]	11 (52%)	4 (31%) (nr = 13)
Total Sample [n = 39]	15 (38%)	4 (21%) (nr = 19)

Note.   Figures are for total sample except where indicated by nr = number reporting.

Table 7.10 gives the data on concerns for which patients considered entering counseling from the most frequently cited to the least cited response. Subjects were asked to check their concerns from a list of 12 problem areas suggested by the literature. Half of each diagnostic group responded, and missing data complicate interpretation. The study groups did not differ significantly in terms of the proportion reporting a given area as problematic. Slightly less than half (46%) of the entire sample reported no concerns. Ranked by percentage for the total sample, the most common reasons for considering counseling were school, romantic, and family concerns, whereas substance abuse, work problems, and anxiety were the least often cited concerns.

DISCUSSION

Our data suggest that the post-treatment adjustment and behavioral functioning of these patients is clearly very positive.Most subjects were high school graduates or better, employed, and not experiencing significant emotional or adjustment problems. A large percentage reported previous counseling experience, a factor that may have been influenced by the long-standing availability of Psychoendocrine services at Children's Hospital. Unfortunately, comparative norms for counseling experience are not available. None of our patients had major psychopathology.

The data on current living situation suggest that these patients maintain close relationships with their parents, because the majority are still at home (67%). This may reflect the sizable number of students in the sample (39%) and/or the economic recession that occurred coincident with this study. On the other hand, their living arrangements may reflect a delay in the

TABLE 7.10

Concerns for Which Patients Considered Entering Counseling

Patient Groups	School	Romantic	Family	Juvenilization/ Teasing	Medical	Sexual	Depression	Work	Anxiety	Drug/ Alcohol Abuse
Isolated GH Deficiency (IGHD) [n = 18]	3 (33%) (nr = 9)	4 (44%) (nr = 9)	5 (56%) (nr = 9)	2 (22%) (nr = 9)	1 (11%) (nr = 9)	1 (11%) (nr = 9)	1 (11%) (nr = 9)	None (nr = 9)	1 (11%) (nr = 9)	None (nr = 9)
Multiple Pituitary Deficits (MPD) [n = 21]	9 (75%) (nr = 12)	7 (58%) (nr = 12)	4 (33%) (nr = 12)	5 (42%) (nr = 12)	5 (42%) (nr = 12)	5 (42%) (nr = 12)	5 (42%) (nr = 12)	4 (33%) (nr = 12)	2 (17%) (nr = 12)	1 (8%) (nr = 12)
Total Sample [n = 39]	12 (57%) (nr = 21)	11 (52%) (nr = 21)	9 (43%) (nr = 21)	7 (33%) (nr = 21)	6 (29%) (nr = 21)	6 (29%) (nr = 21)	6 (29%) (nr = 21)	4 (19%) (nr = 21)	3 (14%) (nr = 21)	1 (5%) (nr = 21)

Note.   Figures are for total sample except where indicated by nr = number reporting.

development of intimate adult sexual relationships. This latter explanation is more applicable to MPD subjects, who left their parental home at an older age than their counterparts in the IGHD group. The MPD patients also reported the lower incidence of marriage, current sexual partnerships, and dating experience. Moreover, the MPD subjects reported a higher incidence of concern about the romantic and sexual areas of their life.

Both the data on juvenilization and self-perception of physical appearance suggest that the MPD patients are more likely to be perceived by others and by themselves as looking inappropriately young. Their perception of how old they look was significantly more discrepant (younger) from their chronological age than that of IGHD patients. The reported prevalence of both past and present juvenilization was greater in the MPD group. It is not clear if past and present experience with juvenilization accounts for the adult living arrangements of many of our patients.

The problems with juvenilization and inappropriately young appearance appear to be worse when there are multiple pituitary hormone deficits. We believe that gonadotropin deficiency is the most likely cause of the phenomena, primarily because of its role in sexual maturation. Furthermore, recent studies indicate that gonadotropin deficient hypopituitary males show marked improvement in virilization and physical appearance following gonadotropin replacement (Clopper et al., 1983).

We conclude that the future looks bright from a behavioral viewpoint for individuals with GH deficiency if treatment begins early in childhood. Recombinant DNA technology should provide sufficient supplies of GH to ensure appropriate height for genetic endowment. Gonadotropin and/or sex steroid therapy can be prescribed, if necessary, to provide age-appropriate sexual maturation. Thus the problems of short stature, juvenilization, inappropriate young appearance, and suboptimal sexual maturation will be minimized—if not corrected.

ACKNOWLEDGMENTS

This work is supported by grants from the Human Growth Foundation and Variety Club Tent #7.

REFERENCES

Abbott, D., Rotnem, D., Genel, M., & Cohen, D. J. (1982). Cognitive and emotional functioning in hypopituitary short statured children. *Schizophrenia Bulletin, 8*, 310–319.

Clopper, R. R., Adelson, J. M., & Money, J. (1976). Postpubertal psychosexual function in male hypopituitarism without hypogonadotropinism after growth hormone therapy. *Journal of Sex Research, 12,* 14–32.

Clopper, R., Mazur, T., MacGillivray, M., Peterson, R., & Voorhess, M. (1983). Data on virilization and erotosexual behavior in male hypogonadotropic hypopituitarism during gonadotropin and androgen treatment. *Journal of Andrology, 4,* 303–311.

Clopper, R. R., Meyer, W. J. III, Udvarhelyi, G. B., Money, J., Aarabi, B., Mulvihill, J. J., & Piasio, M. (1977). Postsurgical IQ and behavioral data on twenty patients with a history of childhood craniopharyngioma. *Psychoneuroendocrinology, 2,* 365–372.

Dorner, S., & Elton, A. (1973). Short, taught and vulnerable. *Special Education, 62,* 12–16.

Drotar, D., Owens, R., & Gotthold, J. (1980). Personality adjustment of children and adolescents with hypopituitarism. *Child Psychiatry and Human Development, 11,* 59–66.

Ehrhardt, A. A., & Meyer-Bahlburg, H. F. L. (1975). Psychological correlates of abnormal pubertal development. *Clinics in Endocrinology and Metabolism, 4,* 207–222.

Holmes, C. S., Hayford, J. T., & Thompson, R. G. (1982). Parents' and teachers' differing views of short children's behavior. *Child care, Health and Development, 8,* 327–336.

Mazur, T., & Clopper, R. (in press). Hypopituitarism: Review of behavioral data. In C. D. G. Brook & D. M. Styne (Eds.), *Advances in Pediatric Endocrinology.* London: Butterworth Scientific.

Meyer-Bahlburg, H. F. L. (1985). Psychosocial management of short stature. In D. Shaffer, A. A. Ehrhardt, & L. L. Greenhill (Eds.), *The clinical guide to child psychiatry* (pp. 110–144). New York: Free Press.

Meyer-Bahlburg, H. F. L., & Aceto, T. (1976, March). *Psychosexual status of adolescents and adults with idiopathic hypopituitarism.* Paper presented at the 33rd annual meeting of the American Psychosomatic Society, Pittsburgh.

Meyer-Bahlburg, H. F. L., Feinman, J. A., MacGillivray, M. H., & Aceto, T. (1978). Growth hormone deficiency, brain development and intelligence. *American Journal of Diseases of Children, 132,* 565–572.

Money, J., Clopper, R., & Menefee, J. (1980). Psychosexual development in postpubertal males with idiopathic panhypopituitarism. *The Journal of Sex Research, 16,* 212–225.

New York State Department of Commerce. (1982). *1980 Census of Population, Summary File Tape 3A: Characteristics of People and Housing.* Albany: New York State Data Center.

Nie, N. H., Hull, C. H., Jenkins, J. G., Steinbrenner, K., & Bent, D. H. (1975). *Statistical package for the social sciences* (2nd ed.). New York: McGraw-Hill.

Pollitt, E., & Money, J. (1964). Studies in the psychology of dwarfism. I. Intelligence quotient and school achievement. *Journal of Pediatrics, 64,* 415–421.

Rosenbloom, A. L., Smith, D. W., & Loeb, D. G. (1966). Scholastic performance of short statured children with hypopituitarism. *Journal of Pediatrics, 69,* 1131–1133.

Rotnem, D., Genel, M., Hintz, R. L., & Cohen, D. J. (1977). Personality development in children with growth hormone deficiency. *Journal of the American Academy of Child Psychiatry, 16,* 412–426.

Weiss, J. O. (1977). Social development of dwarfs. In W. T. Hall & C. L. Young (Eds.), *Genetic disorders: Social service interventions.* Pittsburgh: University of Pittsburgh.

8 Psychosocial Impact of Long-term Growth Hormone Therapy

C. M. Mitchell
Ann J. Johanson
Susan Joyce
University of Virginia,
Charlottesville, Virginia

Samuel Libber
Leslie Plotnick
Claude J. Migeon
Johns Hopkins University School of
Medicine, Baltimore, Maryland

Robert M. Blizzard
University of Virginia School of Medicine,
Charlottesville, Virginia

Human growth hormone (hGH) was first administered intramuscularly to a hypopituitary patient in 1958 (Raben, 1958). In the early 1960s, small numbers of patients were treated with hGH, which was extracted from human cadaveric pituitary glands that were collected and subjected to extraction procedures at a variety of institutions around the world. Because of uncertain supply, concern about misuse and abuse, and in an attempt to supply hGH to needy patients, the National Pituitary Agency was formed in Baltimore, MD. This agency subsequently became the major source of hGH for the treatment of GH deficient patients in this country. Because of limited collection of pituitaries and incomplete extraction, the availability of hGH was restricted. Thus, often only the most severely GH deficient children were treated. Treatment was often late in childhood; the dose was limited; and treatment often was interrupted. Increased pituitary collection rates and improved yield from extraction procedures have resulted in improved treatment opportunities for GH deficient children (Burns, Tanner, Preece, & Cameron, 1981). In the 1960s and early 1970s, restricted hGH supplies resulted in the exclusion of boys from treatment once they reached

5'4", and girls who had reached 5'. More recently GH supplies have permitted earlier treatment at higher doses, usually without interruption and for a longer period of time, so that hGH-treated children are now achieving more normal adult stature. Even so, we have recently been faced with serious interruptions in the supply of pituitary GH, primarily because of a declining number of autopsies in this country.

Children with short stature have been viewed as isolated and prone to gravitate toward younger groups (Rotnem, Cohen, Hintz, & Genel, 1979; Rotnem, Genel, Hintz, & Cohen, 1977). Short-statured children reportedly have poor school performance despite normal intelligence (Abbott, Rotnem, Genel, & Cohen, 1982; Meyer-Bahlburg, 1985). Negative self-concept of children with short stature also has been observed (Kusalic & Fortin, 1975; Rotnem et al., 1977; Rotnem et al., 1979). The majority of these and other studies have relied on projective tests and case studies for data-gathering. Statistical analyses are often lacking or difficult to interpret (Money, Clopper, & Menefee, 1980), and many of the studies have used extremely small sample sizes (Spencer & Raft, 1974). Finally, there have been no systematic studies of affected children *after* treatment was completed (Abbott et al., 1982). Very little is known about the ultimate adjustment to adult life following short stature in childhood (Folstein, Weiss, Mattelman, & Rose, 1981).

In the present study, we present data on a group of GH deficient individuals treated with hGH for various periods of time during the previous 25 years. The aim of the study was to assess the psychosocial impact of growth hormone treatment on these individuals. To achieve this, we attempted to measure success in life as determined by usual standards of scholastic, employment (or equivalent), and social achievement, and assessments of self-esteem.

METHODS

Subjects

Seventy patients with GH deficiency were identified from Pediatric Endocrine Clinic charts at Johns Hopkins University and the University of Virginia. All had received long-term hGH treatment (many years). Of the 70, 58 of the patients (and/or their parents) responded to a request to participate in the study. In 46 cases, both patients and parents responded. Forty-four patients were males and 14 were females. The average age was 26 years ± 6.6 SD (range, 16–46 years). All were Caucasian. The average final height of all patients was 157.5 cm (5'2"). Of the males, 75% achieved a height greater than 152.4 cm (5'). Of the females, 10 of 14 achieved a height less than 147.3 cm (4'10").

Eight patients, all males, had organic hypopituitarism as defined by the presence of a tumor mass or a space-occupying intracranial lesion. Nine patients, seven of them females, had idiopathic panhypopituitarism, Forty-one patients had idiopathic hypopituitarism with growth hormone deficiency, but normal gonadotropin and testosterone or estrogen production (34 males; 7 females). Patients with idiopathic hypopituitarism had not been classified satisfactorily relative to TSH, or ACTH deficiency. Many were thought to have limited ACTH reserve as a result of inadequate response to metopirone, when tested many years previously. Many also had intermittently low thyroxine determinations while on growth hormone treatment. Subsequent studies of GH treatment in a number of patients initially thought to have ACTH and/or TSH deficiency were normal; consequently, specific definition of TSH and/or ACTH deficiency was incomplete.

Measures Used for Psychosocial Evaluation

Basically identical questionnaires (one for the patient and one for the parents) were created to obtain general descriptive information (e.g., age, sex, height, weight, employment status). Educational and employment histories were requested for the patient and the patient's siblings and parents. Open-ended questions were included to assess overall education achievement, satisfaction with employment, benefits and hindrances of short stature, positive and negative impacts of hGH treatment on the patient and family, and initial treatment expectations of patients and parents. In addition, parents were asked to compare the patient's general adjustment and educational/vocational accomplishments to those of his or her siblings.

Questionnaires were sent to the parents and the patient. If no reply was received within 4 weeks, either another questionnaire was sent to the patient, or parents were contacted by phone. Once the questionnaire was returned, a phone interview was arranged to obtain more detailed information and to fill in missing data.

The Tennessee Self-Concept Scale (TSCS) (Fitts, 1965) was also mailed after completion of the questionnaire. This measure has 100 self-statements that the respondent rated on a 5-point scale ("completely false" to "completely true"). These items were summed into 10 dimensions of self-concept: identity, self-satisfaction, behavior, physical self, moral-ethical self, personal self, family self, social self, self-criticism, and global self-concept. Table 8.1 provides additional explanation. The reliabilities (i.e., internal consistency) of each scale were checked and found to be satisfactory.

TABLE 8.1
Tennessee Self Concept Scale Subscales

---

1. *Self-identity:* items measuring "what I *am*"; how the person sees him/herself.
2. *Self-satisfaction:* How the person *feels* about the self he/she perceives; taps expectations and standards.
3. *Behavior:* "what I *do*"; perceptions of actual overt behavior.
4. *Physical self:* perceptions of physical body, health, appearance, skills and sexuality.
5. *Moral-ethical self:* moral worth, religious life, feelings of being a "good" or "bad" person.
6. *Personal self:* personal worth, adequacy as a person.
7. *Family self:* adequacy and worth as family member, in closest circle of associates.
8. *Social self:* adequacy and worth in social interaction with others outside of the family.
9. *Self-criticism:* ten items (taken from the L-scale of the MMPI) that are mildly derogatory which most people admit as being true for them. High scores may denote extreme lack of psychological defenses; low scores, defensiveness to acknowledging negative aspects of one's personality.
10. *Global self-concept:* an overall frame of reference within which a person describes him/herself on a variety of aspects of self-concept (summation of other scales).

Fifty-eight families returned the initial questionnaire. In 12 families only one questionnaire was returned—seven from parents and five from patients. In the other families, both questionnaires were returned. The TSCS questionnaire was returned by 29 patients.

RESULTS

Educational/Vocational Assessment

Table 8.2 presents patients' perceptions of academic performance and the impact of short stature on academic performance. Forty-seven percent rated their performance average. Only 12% of the patients believed their school performance to be less than average and 41% viewed their achievement as above average.

Parents differed from the patients in their perceptions of the influence of short stature on the patient's school performance. Fifty-one percent of the parents believed that stature had a negative impact, but only 20% of the patients had a similar view. Eighteen percent of patients reported a positive effect of short stature. The most frequently cited positive effects were "working harder at school as a result of the short stature," and "receiving special attention for being different." The most commonly mentioned negative factor was "teasing by peers." Parents also noted that an important way that the short stature affected the patients was the lack of full participation in physical education classes and sports.

TABLE 8.2
Educational/Vocational Performance

Variable		
Patient perception of school performance:		
Above average	41%	
Average	47%	
Below average	12%	
Parent perception of impact of short stature on school performance:		
Positive impact	2%	
No impact	47%	
Negative impact	51%	
Patient perception of impact of short stature on school performance:		
Positive impact	18%	
No impact	62%	
Negative impact	20%	
Employment status:		
Full-time	57%	
Part-time	10%	
Unemployed	33%	(50% of this category were students or homemakers not desiring employment outside the home)
Employment satisfaction:		
Satisfied	68%	
Dissatisfied	23%	
No answer	9%	
Impact of short stature on securing employment:		
Positive impact	2%	
No impact	57%	
Negative impact	37%	
Not sure	4%	

Parents also were asked to report on educational achievements of the patients' siblings. No significant differences were reported between the patients and their siblings on such variables as number of educational institutions attended or final degrees obtained.

The responses to questions about status of employment and satisfaction with employment are shown in Table 8.2. The rate of unemployment is higher in this sample than the unemployment rate for the country (16.5% versus 7.5% at the time the study was done). Of those employed, 23% were dissatisfied; however, there are no norms against which to compare this. On the positive side, 68% were satisfied with their employment. Thirty-seven percent believed that short stature was a deterrent (negative impact) in their securing employment. When negative impact was noted, the most frequently cited reasons were relative to physical limitations of size (e.g., "not enough strength," "could not reach things," etc.).

Peer/Social Relations

The two most commonly reported effects of short stature on the relationships with peers were the lack of adequate relationships with peers of the same sex and age, and difficulties with heterosexual social relationships. The two most frequently cited effects within the family were the strain involved in trying to diagnose and treat the growth problem, and the sibling jealousy that was aroused by the special attention the patient frequently received.

The patients' and parents' responses regarding appearance versus chronological age were in agreement. Fifteen percent believed that the patient, as an adult, looked at least 10 years younger than his or her chronological age; 42% believed his or her appearance to be between 5 and 9 years younger than the chronological age; 28% perceived himself or herself to appear 1 to 4 years younger than the chronological age; and 15% believed that the patient looked his or her chronological age.

When asked if the patients were treated as younger than their chronological age when they were children, 60% of the patients said yes and 40% replied no; 80% of the parents, however, reported that such treatment had occurred and 20% said that it had not. When the patients were asked whether they currently were treated as younger than their chronological age, 60% reported that they were not.

Table 8.3 displays basic information about psychosexual development of the patients. At least 78% had dated. Thirty nine percent reported having had sexual intercourse. Thirty-five percent elected not to answer this question. Sixty-one percent did not answer the question regarding a current sexual partner. Of those replying (39%), 22% had a partner.

Of the 52 responding to inquiries regarding marital status, 35 were single and 17 had been married or were currently married. (Two of these 17 were divorced.) Ten of the 17 couples were parents: 8 couples had biologic children; 2 couples had adopted children. There were 17 children among the 10 couples. Psychological differences between the group with natural children and those without children were explored. The following differ-

TABLE 8.3
Psychosexual Development

	Yes	No	No Answer
Ever dated	78%	10%	12%
Ever had sexual intercourse	39	26	35
Currently have a sexual partner	22	17	61
Average age at first date: 16.5 years			
Average age at first sexual intercourse: 23 years			

ences (at a level of at least $p < .05$ — due to low intercorrelations of the variables, univariate t-tests were used) were noted:

1. Short stature was a more negative factor in school performance for the group with children than for those without children.
2. Short stature was a more negative factor in employment for those in this group than for those without children.
3. The expectations for treatment were less well met for this group than for those without children.
4. These patients made more negative statements about themselves on four subscales of the TSCS that did patients who were not parents (self-identity, self-satisfaction, physical self, and overall self-concept). (See Table 8.4.)

Self-Concept

Two measures of self-concept were examined. The TSCS was utilized to compare this treated sample to a "normal stature" group (established norms). Second, on the original questionnaire, the parents compared the patient's overall adjustment to that of his or her siblings. (Demographic characteristics of those returning the TSCS measure were compared to those *not* returning it, in order to detect possible biases in the subsample. No statistical differences were found.)

The scores of the TSCS for the 29 patients were compared to established norms. The patients as a group rated themselves significantly higher in

TABLE 8.4
Comparison of Those With and Without Natural Children

Variable	Direction of Effects	Test Statistics[a]
Short stature effect on academic performance	Lower[b]	$F = 4.41$
Short stature effect on employment	Lower	$F = 5.49$
Treatment expectations met	Lower	$F = 10.86$
Self-identity	Lower	$F = 5.52$
Self-satisfaction	Lower	$F = 6.35$
Physical self	Lower	$F = 9.06$
Overall self-concept	Lower	$F = 6.55$

[a] The F statistic is a ratio of variances; all test statistics are significant at a level of $p < .05$ or less, indicating that the differences that occurred were greater than that expected by chance.

[b] "Lower" signifies that the group with natural children scored lower, or more negatively, than the group without natural children.

self-satisfaction, personal worth, and sociability than the normative group, but significantly lower in their evaluation of physical self and self-criticism. (See Table 8.5.)

Sibling comparisons were explored with the parents. The majority (55%) noted no critical differences in general adjustment or achievements between the patient and his or her siblings of normal stature. Fifteen percent of the parents believed that the patient was better adjusted than the siblings, and 30% believed the siblings were better adjusted than the patient.

Other Findings

Patients and parents were questioned about their *expectations and recommendations* for treatment. Table 8.6 presents the results of these inquiries. Expectations were stated to have been met for 64% of the patients and for 50% of the parents. This was true even though many were treated late, and did not achieve a height that would be considered to be in the normal range. Overwhelmingly, the greatest benefit of the program was the patients' growth; more specifically, many patients and parents noted that the patient was better able to interact in society by being taller. The worst aspects of the treatment were the pain of the injections and inconveniences of the treatment—primarily, traveling to and staying at the treatment site.

Patients and parents were in accord about recommendations of treatment to others. Eighty-five percent recommended the treatment to patients with growth hormone deficiency; 10% recommended it with some qualifications (e.g., not for children who just want to grow faster, i.e., just want short-term effects); 5% said they would *not* recommend treatment to others.

The suggestion that patients and parents most frequently made to physicians currently working with short-statured children was to establish self-

TABLE 8.5
Treated Sample Vs. Normative Group—TSCS

Variable	Direction of Effects	Test Statistics[a]
Self-satisfaction	Higher[b]	$F = 3.84$
Personal self	Higher	$F = 6.86$
Social self	Higher	$F = 4.62$
Physical self	Lower	$F = 5.62$
Self-criticism	Lower	$F = 6.55$

[a]The F statistic is a ratio of variances; all test statistics are significant at a level of $p < .05$ or less, indicating that the differences that occurred were greater than that expected by chance.

[b]"Lower" signifies that the treated sample scored lower, or more negatively, than the norms; "higher," that the treated sample scored higher, or more positively, than the norms.

TABLE 8.6
Expectations of Treatment

	Patients	Parents
Initial expectations:		
None	18%	6%
To grow some	41	47
To attain "normal" height	37	37
Too high	4	10
Were expectations met:		
Yes	64%	50%
Only partially	32	48
Not at all	4	2

help support groups of families and patients with growth problems. The need to downplay the deviance of being short was also emphasized (e.g., have successful short people talk with the patient, in the event the patient does not attain "normal" height). In addition, the patients commented about the impersonal treatment they often received from physicians (e.g., large numbers of physicians in the room; speaking in front of the patients as if they were not there). Finally, they strongly recommended that prospective patients be given realistic expectations about the possibility of growth in response to treatment. (Thirty-seven percent of patients and parents expected patients to reach normal height, but only 4% of patients and 10% of parents felt that their expectations were too high.)

DISCUSSION

Before discussing the findings and implications of this study, some limits of the statistical evaluation need explanation.First, because 12 patients could not be contacted or never responded to the initial questionnaire, it is impossible to know whether a sample bias may exist in the reported data set. Second, in some of the subsample comparisons, there were very small cells. For example, in comparing parents of natural children to those with no children on the TSCS data, one of the cells was made up of only six subjects. Because of this, some of these analyses may be statistically less stable, requiring future elaboration.

Third, in comparing this sample to "normal-stature" (either siblings or normative) groups, few differences were discovered. However, the fact that no differences were found does not necessarily imply that there were no differences between the groups. Perhaps there were differences if compared with unanalyzed groups or if the instruments utilized were not powerful enough to discriminate. Finally, the research design used was a purely post hoc, correlational design and much of the information was retrospective. Unfortunately, perceptions of the past sometimes may be altered by the passage of time. Similarly, there may have been a "halo" effect or perceived pressure to respond positively about the program. Even though the data collectors were psychologists who had not been involved in the treatment originally, the involvement of the therapists (the pediatric endocrinologists) was considerable and could have been conducive to more positive responses than are truly representative. In addition, no implications of causality can be drawn from any correlational design. One can merely report the occurrence or non-occurrence of relationships between variables without inferring the direction of a causal relationship between them. However, when these cautions are considered, one can explore some of the broader implications of this study.

Before focusing on the three research areas of primary interest, two slightly tangential findings should be addressed. First, correlations between parent and patient reports generally were quite high, with two notable exceptions. The parents reported that short stature had a more negative impact on the patients in school than was reported by the patients. The parents also reported more often than did the patients that the patient as a child had been treated more in accord with his or her "height age" rather than in accord with the chronological age. In all other areas where both the patient and the parent were asked the same questions, there was a high degree of agreement. It is possible that the observed disagreement was simply an artifact of retrospective data. Perhaps the parents always remember the past differently than the child does. Another interpretation might be that the parents were more aware of certain aspects than was the patient or perhaps the parents interpreted events more negatively than did the patient. Although the true interpretation cannot be discerned from this study, there is support in the literature for the latter explanation. Parents (especially mothers) have reported early negative views of their short-statured children (e.g., sickly, vulnerable). In addition, there is a tendency to doubt one's parenting competence (Rotnem et al., 1979), which may have resulted in parental misattributions of events as more serious than the child perceived them.

The second area of tangential findings which was unexpected was the consistently negative reports about themselves from the eight people with natural children. Possibly the patients without children were inflating their self-images as a defense. Perhaps those who have families, being more settled and secure, no longer felt the need to "kid" themselves or others. Or possibly, this subgroup felt negatively about themselves all along, and one way that they attempted to improve their image was through raising a family. Because this is a correlational design, neither of these interpretations can be verified. However, Folstein et al. (1981) reported a similar finding with untreated short-statured adults, as well. Further research and more longitudinal designs are needed to clarify this.

The first research question examined the patients' educational and vocational accomplishments. An important comparison was with the patients' siblings. Overall, there were no consistent differences in the educational achievement or job histories of the two groups. In addition, patients, in general, reported average and above average school performance as well as satisfaction with their employment situations. These findings offer further support to those studies that refute the lack of successful adjustment in short-statured people (Pollitt & Money, 1968; Rosenbloom, Smith, & Loeb, 1966).

The second area this study *specifically* addressed was that of peer interaction. Clearly, this sample found social relationships to have been troublesome, even in retrospect. Because many in this sample were still short in stature, despite intervention, this is an area that deserves much attention and sensitivity from professionals. It is also possible that differences might be anticipated in a group in whom more normal stature is achieved.

The third research area was the comparison of the treated sample to a normative group with respect to self-concept. Not surprisingly, this sample felt less positive about themselves physically, a finding that has been reported by others (Kusalic & Fortin, 1975). However, it appears that this sample fared well as a group, for they seemed to be more self-satisfied and to feel better about themselves. There were no other differences regarding aspects of self-concept between the sample and the normative group. Because others have reported more negative self-concepts (Cronbach, 1951), these results may be unexpected. However, the fact that this group also scored *lower* on self-criticism in general may be an important key in interpreting the information. Usually, scoring low on self-criticism is interpreted as evidence of defensiveness against saying negative things about oneself. Indeed, Rotnem et al. (1979) reported that short-statured populations tend to use denial in dealing with problems. Denial of problems, coupled with "accentuating the positive," might explain these results. Thus, it is possible that defensiveness may have resulted in inflated reports of positive self-perceptions.

An alternative explanation is that this sample has honestly learned to regard themselves more kindly and to exhibit more compassion toward themselves than occurs in unafflicted people. In fact, there is some support for this perspective in this data set. Many parents and patients noted compassion toward self and others as an important positive effect of short stature.

It is clear that the expectations of many patients and parents often were not met. The same effect has been noted (Rotnem et al., 1977) by others among growth hormone treated individuals. Physicians must be extremely realistic in explaining to patients the benefits of treatment. For example, the physician should emphasize that the long-range goal is to "catch up in height," but the short-term goal is to "not fall further behind."

CONCLUSION

This study has demonstrated that being short results in some problems of adjustment.However, many patients in this group did very well with respect to scholastic achievement, employment, and social and family adjustment. In retrospect, hGH treatment was perceived as a positive factor by most of these patients. In this sample, earlier treatment might have been more successful in minimizing the handicaps of being short. Thus, further studies comparing the results in this sample with groups treated earlier will be helpful in the treatment of short stature. Regardless of future findings, this study has shown that hGH treatment can be instituted in late childhood with positive psychological as well as physical benefits to many patients.

ACKNOWLEDGMENTS

These studies were supported by the Human Growth Foundation and NIH General Clinical Research Centers Grant RR-847.

REFERENCES

Abbott, D., Rotnem, D., Genel, M., & Cohen, D. J. (1982). Cognitive and emotional functioning in hypopituitary short-statured children. *Schizophrenia Bulletin, 8,* 310.

Burns, E. C., Tanner, J. M., Preece, M. A., & Cameron, N. (1981). Final height and pubertal development in 55 children with idiopathic growth hormone deficiency treated for between 2 and 15 years with human growth hormone. *European Journal of Pediatrics, 137,* 155.

Cronbach, L. J. (1951). Coefficient alpha and the internal structure of tests. *Psychometrika, 16,* 297.

Fitts, W. H. (1965). *Tennessee Self Concept Scale.* Los Angeles: Western Psychological Services.

Folstein, S. E., Weiss, O. J., Mattelman, F., & Rose, D. J. (1981). Impairment, psychiatric symptoms and handicap in dwarfs. *The Johns Hopkins Medical Journal, 148,* 273.

Kusalic, M., & Fortin, C. (1975). Growth hormone treatment in hypopituitary dwarfs: Longitudinal psychological effects. *Canadian Psychiatric Association Journal, 20,* 325-331.

Meyer-Bahlburg, H. F. L. (1985). Psychosocial management of short stature. In D. Shaffer, A. A. Ehrhardt, & L. L. Greenhill (Eds.), *The clinical guide to child psychiatry* (pp. 110-135). New York: Free Press.

Money, J., Clopper, R., & Menefee, J. (1980). Psychosexual development in postpubertal males with idiopathic panhypopituitarism. *Journal of Sex Research, 16,* 212.

Pollitt, E., & Money, J. (1964). Studies in the psychology of dwarfism I. Intelligence quotient and school achievement. *Journal of Pediatrics, 64,* 415–421.

Raben, M. S. (1958). Treatment of a pituitary dwarf with human growth hormone. *Journal of Clinical Endocrinology and Metabolism, 18,* 901.

Rosenbloom, A. L., Smith, D. W., & Loeb, D. G. (1966). Scholastic performance of short-statured children with hypopituitarism. *Journal of Pediatrics, 69,* 1131–1133.

Rotnem, D., Cohen, D. J., Hintz, R., & Genel, M. (1979). Psychological sequelae of relative "treatment failure" for children receiving human growth hormone replacement. *Journal of the American Academy of Child Psychiatry, 18,* 505–520.

Rotnem, D., Genel, M., Hintz, R., & Cohen, D. J. (1977). Personality development in children with growth hormone deficiency. *Journal of the American Academy of Child Psychiatry, 16,* 412.

Spencer, R. F., & Raft, D. D. (1974). Adaptation and defenses in hypopituitary dwarfs. *Psychosomatics, 15,* 35.

9 Cognitive Development in Turner Syndrome

Hans-Christoph Steinhausen
Free University of Berlin, West Germany

Jacqui Smith
Max-Planck-Institute for Human Development and Education, Berlin, West Germany

The syndrome of webbed neck, cubitus valgus, short stature, and sexual infantilism was described in seven patients by Turner in 1938. Today, this syndrome is associated with his name although in 1930, the German pediatrician Ullrich reported the cardinal features of the syndrome. Parts of the syndrome, particularly dysgenesis of the gonads, had even been noted in the mid-18th century. In 1959, the new technique of chromosome analysis enabled Ford, Polani, Jones, de Almeida, and Briggs to show that a 45,XO Karyotype was the cause of the disorder. Later Fraccaro, Ikkes, Lindsten, and Kaijser (1960) detected that other chromosomal anomalies involving the second X chromosome could be associated with the syndrome. In these cases, there is either 45,X mosaicism or various other abnormalities of the X chromosome pair. The incidence of the syndrome is estimated at 1 per 3,000 female births, with but a small proportion of approximately 5% of those conceived surviving to birth.

The main clinical features of short stature and ovarian dysgenesis leading to primary amenorrhea are present in all cases of Turner's syndrome (TS). Cardiovascular malformations are less common than renal malformations and the webbed neck occurs in 50% of patients. Further symptoms include edema of the dorsum of the hands and feet recognizable at birth. In childhood besides webbing of the neck, a low posterior hairline, small mandible, prominent ears, epicanthal folds, high arched palate, a broad chest, cubitus valgus, and hyperconvex fingernails contribute to clinical manifestations. Sexual maturation fails to occur at the usual time during puberty. In the XO/XX mosaic the abnormalities are fewer and attenuated.

An understanding of the intellectual and cognitive capabilities of females with TS is of considerable importance, especially for the study of the etiology of individual and possible sex-related differences. If, for example, an aspect of cognitive development or cognitive functioning is thought to be primarily genetically sex-linked, then detailed descriptions of the relevant phenotypic variations among individuals with sex chromosome anomalies, like the TS females, can shed light on the viability of such theories. The aim of this chapter is to set TS research on cognitive functioning in historical perspective and to point to future research directions.

Because sex chromosome abnormalities are often connected with mental retardation, the first psychological investigations of TS patients addressed this issue. These early studies focused on the assessment of general intellectual functioning and are reviewed in the first section of this chapter. It soon was recognized that specific cognitive and neuropsychological deficits, rather than a generalized lower level of intellectual functioning, were characteristic of TS patients. These findings are discussed in the second section of the chapter. From a different perspective, some researchers have addressed the question of whether or not certain somatic features of the syndrome were connected with intelligence; these findings are reviewed in the third section of this chapter. Finally, the theoretical implications of this research are discussed.

GENERAL INTELLECTUAL FUNCTIONING

Early clinical reports suggested that all patients with TS were mentally retarded or of below average intelligence (e.g., Haddad & Wilkins, 1959). However, it soon was suggested that only the incidence of retardation among TS females was greater than among the general population. In an early summary of the literature, Ferguson-Smith (1965) found 19 cases of "severe mental defect" among 287 females. This representation may, however, have been inflated because these surveys searched for females with webbed necks and/or chromosome abnormalities in institutions for the severely retarded. In the early 1960s, two systematic surveys were performed in the Netherlands showing that the incidence of TS persons among retardates was no greater than the incidence of TS persons in the general population (see Garron & Vander Stoep, 1969).

This conclusion, however, has been challenged by a more recent survey performed in Denmark. Nielsen and Sillesen (1983) studied the incidence of TS in all girls born between 1955 and 1966 living in one of the 10 institutions for mentally retarded in Denmark. Among 2,880 girls, they identified 8 TS patients, leading to an incidence rate of 1:360 or 3.3 per thousand, which is 9 times higher than the expected rate of 1:3,000. Other studies (de la Chapelle, 1963; Harms, 1967; MacLean et al., 1962; Yanagisawa & Shuto, 1970), based on large surveys of mentally retarded persons in different countries, found incidence rates of 0.5 to 2 per thousand. Nielsen and Sillesen (1983) concluded that the incidence of TS among retardates must be somewhere between 2 and 3 per thousand, which actually is 8 to 10 times higher than expected. On the other hand, these same authors (who have so far reported on the largest series of TS persons) found only 5 among 82 subjects with subnormal intellectual functioning. This frequency is in accordance with the expected incidence in the general population, namely 3%. So far, no interpretation has been offered for the apparent overrepresentation of TS persons in institutions for mentally retarded. Perhaps this might be primarily due to their physical appearance.

Unfortunately, Nielsen and Sillesen (1983) did not report results of intelligence testing. However, there are a number of surveys published in the literature. The main findings are collected in Table 9.1. Based on a first sample reported by Shaffer (1962), Money and co-workers were the first to systematically assess a large series of patients and to report their findings in a number of papers. Their latest report included a final sample of 46 TS females (Money & Granoff, 1965). Of the 46 females, eight had full-scale IQs below 80, and only one above 119. Although the number of low IQs was overrepresented in their sample, only 5 cases had verbal IQs below 80, whereas 16 cases had performance IQs below 80. Thus, it became apparent that the increased incidence of moderate retardation, as judged from full-scale IQs, was entirely attributable to poor performance IQs. Similar observations have also been made in other samples (Garron, Molander, Cronholm, & Lindsten, 1973; Lindsten, 1963). The finding that TS is not associated with a general lowering of verbal intelligence has been replicated repeatedly. Studies performed in the United States (Bender, Puck, Salbenblatt, & Robinson, 1984; Chen et al., 1981; Garron, 1977; Netley & Rovet, 1982; Rovet & Netley, 1980), Sweden (Garron et al., 1973; Lindsten, 1963), Austria (Haselbacher, 1975), and Germany (Steinhausen, Ehrhardt, & Grisanti, 1978) have shown, with only one exception (Bender et al., 1984), that performance IQ in TS females is significantly lower than verbal IQ (see Table 9.1). This verbal/performance IQ discrepancy is due to poor functioning in those subtests related to perceptual organization. The observation of this phenomenon, which is independent of age (Garron, 1977), stimulated further studies investigating specific cognitive functions in TS females.

TABLE 9.1
General Intelligence Test Findings in TS Patients

Author		N	Age (years) Mean	FIQ[a] Mean	SD	VIQ[b] Mean	SD	PIQ[c] Mean	SD
1. Money & Granoff	1966	46		96.2	15.8	105.6	18.2	86.4	13.6
2. Haselbacher	1975	23		94.6		100.7		89.5	
3. Garron	1977	37 Adults	19	97.2	12.4	104.2	14.1	88.4	12.8
		30 Children	10	91.9	18.3	95.6	17.3	89.1	17.6
4. Steinhausen et al.	1978	8	16.3	85.2	20.9	94.1	20.9	79.8	20.1
5. Rovet & Netley	1980	31	16.8			99.2	10.8	87.0	11.3
6. Chen et al.	1981	17				98.4	14.4	88.1	11.1
7. Netley & Rovet	1982	35	14			99.9	12.5	88.0	12.8
8. Bender et al.	1984	16	12.2	94.6	19.1	95.8	19.2	93.6	18.4

[a]FIQ = Full IQ   [b]VIQ = Verbal IQ   [c]PIQ = Performance IQ

COGNITIVE AND
NEUROPSYCHOLOGICAL FUNCTIONS

The main findings related to specific cognitive functions are collected in Table 9.2. First, analysis of the Wechsler Scales of Intelligence (Money, 1963, 1964; Shaffer, 1962) using Cohen's (1957, 1959) factor scores indicated that the Perceptual Organization factor (based on Block Design and Object Assembly) and the Freedom from Distractibility factor (based on Digit Span and Arithmetic) were both significantly lower than the Verbal Comprehension factor (based on Information, Comprehension, Similarities, and Vocabulary). These findings were replicated in another series by Garron (1977). Further studies of cognitive functioning in TS persons have focused on difficulties with visuo-spatial tasks. In fact, poor direction sense (Alexander, Walker, & Money, 1964), poor performance in tests of visual-constructional abilities (Alexander, Ehrhardt, & Money, 1966), field dependency (Nyborg & Nielsen, 1977) and other areas of spatial perception and organization (Rovet & Netley, 1980; Silbert, Wolff, & Lilienthal, 1977; Steinhausen et al., 1978; Waber, 1979) have given rise to the speculation that the right cerebral hemisphere is selectively impaired in females with TS (Kolb & Heaton, 1975; Silbert et al., 1977; Steffen, Heinrich, & Kratzer, 1978). It was also suggested by Money (1973) that the parietal area was particularly affected.

However, TS females show comparable difficulties with tasks in which performance is not thought to depend primarily on the function of the right hemisphere or the parietal area, as may also be seen from Table 9.2. For instance, poor performance on word fluency reported in different studies

(Money & Alexander, 1966; Waber, 1979), reasoning and numerical ability (Steinhausen et al., 1978), Arithmetic and Digit Span (Shaffer 1962), and Rhythmic Memory (Silbert et al., 1977) indicate that TS females also have some language and memory difficulties. It is difficult to interpret these data, however, because the attention problems of TS females (i.e., as indicated by their low score on Cohen's Freedom from Distractibility Factor) would interfere with their ability to hold long sequences in short-term memory. It has even been argued on the basis of dichotic listening performance, where TS subjects were less likely than controls to show the usual right ear advantage, that these persons have either a reversal or an absence of left hemispheric specialization for verbal processing (Netley & Rovet, 1982). The suggestion here is that, unlike normal individuals, verbal functions in TS subjects are widely represented in both hemispheres (Rovet & Netley, 1982). Findings from the most extensive neuropsychological investigations so far reported (Pennington et al., 1982; Waber, 1979) indicate that information processing deficits associated with TS, although predominantly due to right hemisphere dysfunctioning, are also bilateral and probably affect performance on a wide variety of tasks including certain visuo-spatial tasks.

Clinical support of this view comes from different sources. It is well known from cognitive deficits in children suffering from Minimal Cerebral Dysfunction that early alterations in brain development have generalized rather than localized effects on brain functions (Rie & Rie, 1980). The same conclusion may also be derived from the theory of brain development proposed by the Russian neuropsychologist Luria (1973). Waber (1979), discussing her own findings, has also advocated the interpretation that many of the cognitive deficits in TS females represent an immature pattern of performance. For example, reduced memory capacity and word fluency, poor automatization, and inability to perceive right and left on another person might reflect, in part, a maturational lag where the performance of TS persons qualitatively resembles that of prepubertal children. This question could only be resolved by longitudinal follow-up.

THE RELATIONSHIP BETWEEN
SOMATIC FEATURES OF TS AND INTELLIGENCE

Another line of research in TS has addressed the relationship between certain abnormal physical characteristics and intelligence in general.For example, in early studies there were some suggestions that retardation was more prevalent among persons with a 45,X karyotype and the webbing of the neck that is associated with this karyotype. Only two studies have included sufficiently large samples to adequately reanalyze this relationship.

TABLE 9.2
Cognitive and Neuropsychological Functions

Author		N	Measures	Type of Process	Findings
1. Money	1964	38	Cohen Factors of the WISC/WAIS	Verbal comprehension, perceptual organization, freedom from distractibility	Perceptual organization and freedom from distractibility significantly lower than verbal comprehension
2. Alexander et al.	1964	13	Road Map Test	Right-left direction sense	Poor direction sense
3. Alexander & Money	1965	17	Gates Reading Survey	Reading ability	No deficiency
4. Money & Alexander	1966	16	Primary Mental Abilities Test	Verbal-meaning, space, reasoning, number, word-fluency	Space and word-fluency scores significantly below the normal level
5. Alexander et al.	1966	18	Benton Visual Retention Test, Bender Visual Motor Gestalt Test, Human Figure Drawing	Visual-constructional ability	Group performance below the level of the general population on all three tests
6. Garron	1977	67	Cohen Factors of the WISC/WAIS	Verbal comprehension, perceptual organization, freedom from distractibility	Perceptual organization and freedom from distractibility significantly lower than verbal comprehension
7. Nyborg & Nielsen	1977	45	Rod-and-Frame Test	Field dependence	Extremely field dependent due to perceptual instability
8. Silbert et al.	1977	13	Neuropsychological Battery	Spatial perception and organization, Sensory-motor sequencing, automatization	Poorer performance on tests of spatial ability and serial processing

(Continued)

TABLE 9.2
(Continued)
Cognitive and Neuropsychological Functions

Author	N	Measures	Type of Process	Findings
9. Steinhausen et al. 1978	8	Primary Mental Abilities Test (PSB)	Verbal-meaning, reasoning, word-fluency, space, numerical ability, perceptual speed	Reasoning, space, numerical ability, and perceptual speed scores significantly below the normal level
10. Steffen et al. 1978	18	Neuropsychological Battery	Language and speech, manual skills, visual abilities, visual and acoustic memory, dichotic listening, right-left-orientation, body image	Right hemispheric maturational lag in only half of the persons
11. Waber 1979	11	Neuropsychological Battery	Dichotic listening, finger tapping, spatial ability, visual memory, right-left-orientation, direction sense, word fluency, visuo-motor co-ordination	Poorer performance on word fluency, perception of left and right, visuo-motor co-ordination, visual memory and motor learning than normal controls
12. Rovet & Netley 1980	31	Mental Rotation Task	Spatial task	Poorer performance in TS compared to normal controls
13. Chen et al. 1981	20	WISC subtests, Bender Visual Motor Gestalt Test	Perceptual organization, visuo-motor functioning	16/20 had space-form perceptual deficits, 13/20 had visual motor deficits
14. Bender et al. 1984	9	Spatial Relations Test, Visual-Motor-Integration, Bender Gestalt, Human Figure Drawing	Perceptual organization	Performance of the nonmosaic group lower than controls

117

With regard to karyotype, neither the series of Money and Granoff (1965) nor the even larger series of Garron (1977) found any differences in IQs between 45,X probands and non-45,X probands. As Table 9.3 shows, these findings have been obtained for both children and adults. Similarly, in a recent study based on a smaller sample Bender et al. (1984) found no obvious developmental differences between TS girls with 45,X genotype and TS girls with 45,X variant, respectively. Thus, there is no evidence that IQ or the specific intellectual skills differ in these subgroups of TS females.

Similarly, the two studies by Garron (1977) and Money and Granoff (1965) and a recent study by Bender et al. (1984) have addressed the question whether any single somatic malformation or a number of stigmata are associated with lower intelligence. All three studies found no evidence that webbing of the neck or any other somatic stigma was correlated with an increase of mental retardation. It is noteworthy that even cardiac defects in TS persons are not associated with lowered levels of intelligence, unlike some other congenital heart defects where cyanosis leads to lowered intelligence levels (Linde, Rasoff, & Dunn, 1970).

There are conflicting findings with regard to the impact of mosaicism in which the abnormal cell line constitutes a minority of the entire cell population in TS females. Chen, Faigenbaum, & Weiss (1981) found no differences of verbal IQ, performance IQ, and full-scale IQ between apparently nonmosaic and mosaic types of TS. Although nothing is known about the selectivity of these findings, subsequent observations by Bender et al. (1984) and Pennington et al. (1982) were based on unselected cases identified through chromosome screening at birth. Here, mosaicism was associated with normal early development and intelligence, whereas the nonmosaic group had a moderately decreased full-scale and performance IQ. However, the representativeness of this sample may also be questioned

TABLE 9.3
The Relationship between Karyotype and Intelligence in TS Patients

		Age (years)	FIQ		VIQ		PIQ	
	N	Mean	Mean	SD	Mean	SD	Mean	SD
Adults								
-45,X probands	21	19.3	95.9	13.5	104.2	13.7	85.6	12.5
-Non-45,X probands	16	19.5	98.9	13.5	103.9	15.0	92.1	12.5
Children								
-45,X probands	21	10.1	94.2	16.1	97.4	16.6	91.6	14.6
-Non-45,X probands	9	10.4	86.4	22.7	91.4	19.2	83.3	23.3

Note.   From Garron, 1977.

because a considerable number of cases could not be followed up after birth. Thus, the question seems to be open whether or not cognitive patterns associated in mosaicism in TS are the same as or different from the cognitive patterns associated with the pure 45,X karyotype or its variants.

IMPLICATIONS AND CONCLUSIONS

Individuals with TS and other naturally occurring errors in sexual development enable researchers to study the impact of sex-linked genes, in addition to autosomal genes acting through hormonal control on brain development in humans. This model allows researchers to study biological determinants of behavioral sex differences that cannot be obtained from a comparison of normal males and females in whom biological and environmental determinants of sex behavior are inextricably linked.

The study of intellectual and cognitive functions in TS females has revealed the following major findings. The incidence of TS persons in institutions for mentally retarded is higher than expected while at the same time the proportion of subnormal intellectual functioning individuals with TS is not greater than in the general population. Although TS persons show a similar distribution of full-scale intelligence like the normal population, they are characterized by a number of specific cognitive and neuropsychological deficits. Deficiencies in the area of visuo-spatial processes are pronounced. As has been shown previously, however, the pattern of neuropsychological dysfunctions is even wider. Moreover, there is no evidence that only certain cortical areas like the parietal lobe or the right hemisphere are affected. Finally, these cognitive deficits are independent of culture, age, karyotype, and/or somatic stigmata.

To date, these data have primarily been examined in the context of theories about the linkage between the X chromosome and spatial ability. In general, females tend to score higher on measures related to verbal ability, whereas males tend to score higher on those related to spatial ability (Maccoby & Jacklin, 1974). The data from TS females appear to support the idea that spatial ability might be genetically sex-linked. Several theories have been advanced to account for the mechanisms of this linkage: for example, that the TS spatial deficit is (a) due to an improper hormonal balance (Peterson, 1976); or (b) an abnormality in growth rate (e.g., Rovet & Netley, 1982; Waber, 1976).

Before firm conclusions can be reached, however, it would seem important to trace the developmental impact of TS on cognitive functioning. In normal individuals, sex differences in visual-spatial versus verbal ability are most apparent following the onset of puberty (Maccoby & Jacklin, 1974).

Longitudinal assessment of TS females on cognitive tasks would help to clarify the nature of their spatial/verbal discrepancy. Most studies so far have combined performance scores from subjects aged 8 to 20 years. Moreover, longitudinal within-subject comparison avoids the extensive methodological problems associated with controlling for between-subject variation in circumstances associated with TS research (e.g., institutionalized vs. non-institutionalized).

Some initial longitudinal data reported by Pennington and Smith (1983) suggested that the TS spatial deficits were present from age 4. It would be important to determine, however, whether this deficit is developmentally stable. Longitudinal data of this type would not only be a significant contribution to cognitive theory but would also be of interest in the context of behavioral genetics as a test of the ideas that genes turn on and off during development (Plomin, 1983).

REFERENCES

Alexander, D., Ehrhardt, A. A., & Money, J. (1966). Defective figure drawing, geometric and human in Turner's syndrome. *Journal of Nervous and Mental Diseases, 142,* 161–167.

Alexander, D., & Money J. (1965). Reading ability, object constancy and Turner's syndrome. *Perceptual and Motor Skills, 20,* 981–984.

Alexander, D., & Money, J. (1966). Turner's syndrome and Gerstmann's syndrome: Neuropsychologic comparisons. *Neuropsychologica, 4,* 265–273.

Alexander, D., Walker, H. T., & Money, J. (1964). Studies in direction sense I. Turner's syndrome. *Archives of General Psychiatry, 10,* 337–339.

Bender, B., Puck, M., Salbenblatt, J., & Robinson, A. (1984). Cognitive development of unselected girls with complete and partial X monosomy. *Pediatrics, 73,* 175–182.

Chen, H., Faigenbaum, D., & Weiss, H. (1981). Psychological aspects of patients with the Ullrich-Turner-Syndrome. *American Journal of Medical Genetics, 8,* 191–203.

Cohen, J. A. (1957). A factor-analytically based rational for the Wechsler Adult Intelligence Scale. *Journal of Consulting Psychology, 21,* 451–457.

Cohen, J. A. (1959). The factorial structure of the WISC at ages 7-6, 10-6, and 13-6. *Journal of Consulting Psychology, 23,* 285–289.

de la Chapelle, A. (1963). Sex chromosome abnormalities among the mentally defective in Finland. *Journal of Mental Deficiency Research, 7,* 129–146.

Ferguson-Smith, M. A. (1965). Karyotype-phenotype correlations in gonadal dysgenesis and their bearing on the pathogenesis of malformations. *Journal of Medical Genetics, 2,* 142–155.

Ford, C. E., Polani, P. E., Jones, K. W., de Almeida, J. C., & Briggs, J. H. (1959). A sex-chromosome anomaly in a case of gonadal dysgenesis (Turner's syndrome). *Lancet I,* 711–713.

Fraccaro, M., Ikkes, D., Lindsten, J., & Kaijser, K. (1960). A new type of chromosomal abnormality in gonadal dysgenesis. *Lancet II,* 1144–1145.

Garron, D. C. (1977). Intelligence among persons with Turner's syndrome. *Behavior Genetics, 7,* 105–127.

Garron, D. C., Molander, L., Cronholm, B., & Lindsten, J. (1973). An explanation of the apparently increased incidence of moderate mental retardation in Turner's syndrome. *Behavior Genetics, 3,* 35–43.

Garron, D. C., & Vander Stoep, L. R. (1969). Personality and intelligence in Turner's syndrome. A critical review. *Archives of General Psychiatry, 21,* 339–346.

Haddad, H. M., & Wilkins, L. (1959). Congenital anomalies associated with gonadal aplasia: Review of 55 cases. *Pediatrics, 23,* 885–902.

Harms, S. (1967). Anomalien der Geschlechschromosomenzahl (XXX- und XO-Zustand) bei Hamburger Hilfsschülern. *Pädiatrie Pädologie, 3,* 34–52.

Haselbacher, L. (1975). Intelligenz- und Persönlichkeitsdiagnostik bei der Gonadendysgenesie (Turner-Syndrom). *Pädiatrie Pädologie, 10,* 244–252.

Kolb, J. E., & Heaton, R. K. (1975). Lateralized neurologic deficits and psychopathology in a Turner syndrome patient. *Archives of General Psychiatry, 32,* 1198–1200.

Linde, L. M., Rasof, B., & Dunn, O. J. (1970). Longitudinal studies of intellectual and behavioral development in children with congenital heart disease. *Acta Paediatrica Scandinavia, 59,* 169–176.

Lindsten, J. (1963). *The nature and origin of X chromosome aberrations in Turner's syndrome.* Stockholm: Almquist & Wiksell.

Luria, A. R. (1973). *The working brain.* New York: Basic Books.

Maccoby, E. E., Jacklin, C. N. (1974). *The psychology of sex differences.* Stanford, CA: Stanford University Press.

MacLean, N., Mitchell, I. M., Harnden, D. G., Williams, J., Jacobs, P. A., Buckbon, K. A., Baikie, A. G., Court Brown, W. M., McBride, J. A., Strong, J. A., Close, H. G., & Jones, D. C. (1962). A survey of sex-chromosome abnormalities among 4514 mental defectives. *Lancet, I,* 293–296.

Money, J. (1963). Cytogenetic and psychosexual incognities with a note on space-form blindness. *American Journal of Psychiatry, 119,* 820–827.

Money, J. (1964). Two cytogenetic syndromes: psychologic comparisons. 1. Intelligence and specific-factor quotients. *Journal of Psychiatric Research, 2,* 223–231.

Money, J. (1973). Turner's syndrome and parietal lobe functions. *Cortex, 9,* 385–393.

Money, J., & Alexander, D. (1966). Turner's syndrome: Further demonstration of the presence of specific cognitional deficiencies. *Journal of Medical Genetics, 3,* 47–48.

Money, J., & Granoff, D. (1965). IQ and the somatic stigmata of Turner's syndrome. *American Journal of Mental Deficiency, 70,* 69–77.

Netley, C., & Rovet, J. (1982). A typical hemispheric lateralization in Turner syndrome subjects. *Cortex, 18,* 377–384.

Nielsen, J., & Sillesen, I. (1983). *Das Turner-Syndrom. Beobachtungen an 155 dänischen Mädchen, geboren zwischen 1955 und 1966.* Stuttgart: Enke.

Nyborg, H., & Nielsen, J. (1977). Sex chromosome abnormalities and cognitive performance: III. Field dependence, frame dependence, and failing development of perceptual stability in girls with Turner's syndrome. *Journal of Psychology, 96,* 205–212.

Pennington, B. F., Bender, B., Puck, M., Salbenblatt, J., & Robinson, A. (1982). Learning disabilities in children with sex chromosome anomalies. *Child Development, 53,* 1182–1192.

Pennington, B. F., Smith, S. D. (1983). Genetic influences on learning disabilities and speech and language disorders. *Child Development, 54,* 369–387.

Peterson, A. C. (1976). Physical androgyny and cognitive functioning in adolescence. *Developmental Psychology, 12,* 524–534.

Plomin, R. (1983). Developmental behavioral genetics. *Child Development, 54,* 253–259.

Rie, H. E., & Rie, E. D. (1980). *Handbook of minimal brain dysfunctions.* New York: Wiley.

Rovet, J., & Netley, C. (1980). The mental rotation task performance of Turner syndrome subjects. *Behavior Genetics, 10,* 437–443.

Rovet, J., & Netley, C. (1982). Processing deficits in Turner's syndrome. *Developmental Psychology, 10,* 77-94.

Shaffer, J. W. (1962). A specific cognitive deficit observed in gonadal aplasia (Turner's syndrome). *Journal of Clinical Psychology, 28,* 403-408.

Silbert, A., Wolff, R. H., & Lilienthal, J. (1977). Spatial and temporal processing in patients with Turner's syndrome. *Behavior Genetics, 7,* 11-21.

Steffen, H., Heinrich, U., & Kratzer, W. (1978). Raumorientierungsstörung und Körperschemairritation bei Turner Syndrom Patienten. *Zeitschrift für Kinder- und Jugendpsychiatrie, 5,* 131-141.

Steinhausen, H. C., Ehrhardt, A. A., & Grisanti, G. C. (1978). Die Beziehung von fötalen Geschlechtshormonen und kognitiver Entwicklung: Studien an Patienten mit adrenogenitalem Syndrom und Turner Syndrom. *Medizinike Psychologie, 4,* 153-163.

Turner, H. H. (1938). A syndrome of infantilism, congenital webbed neck, and cubitus valgus. *Endocrinology, 23,* 566-574.

Ullrich, O. (1930). Ober typische Kombinationsbilder multipler Abartung. *Zeitschrift für Kinderheilkunde, 49,* 271-276.

Waber, D. P. (1976). Sex differences in cognition: A function of maturation rate? *Science, 192,* 572-574.

Waber, D. P. (1979). Neuropsychological aspects of Turner's syndrome. *Developmental Medicine and Child Neurology, 21,* 58-70.

Yanagisawa, S., & Shuto, T. (1970). Sex chromatin survey among mentally retarded children in Japan. *Journal of Mental Deficiency Research, 14,* 254-262.

10 Turner Syndrome versus Constitutional Short Stature: Psychopathology and Reactions to Height

Jennifer Downey

Anke A. Ehrhardt

Rhoda Gruen

Akira Morishima

Jennifer Bell

New York State Psychiatric Institute and Columbia University College of Physicians and Surgeons

The most constant physical feature of women with Turner syndrome (TS) is that they are short.However, women with TS usually have other problems as well—estrogen deficiency, pubertal delay, and infertility. All of these are secondary to a chromosomal abnormality. Although TS in children has been quite carefully studied, the effects of these difficulties on adult psychological adjustment are largely unknown.

Early psychological studies of girls with TS focused on three observations. First, in comparison with controls, TS girls tend to have lower performance IQ scores because of problems with space-form perception and visual-motor coordination (Money, 1973; Waber, 1979). Second, Ehrhardt, Greenberg, and Money (1970) found that girls with TS were as feminine as controls in play behavior and interests. Finally, several investigators noted a personality style described by Money and Mittenthal (1970) as "an inertia of emotional arousal . . . compliance, phlegmatism, stolidity, equability, acceptance, resignedness, slowness in asserting initiative, and tolerance of personal adversity" (p. 54).

Recent studies have raised some questions about adjustment and lack of personality pathology in girls with TS during adolescence and young adulthood. For instance, Taipale (1979) studied 49 girls 9 years of age or older and found them socially withdrawn, without friends of either sex. Their self-confidence as females faltered when puberty did not take place at a time comparable to age-mates. Nielsen, Nyborg, & Dahl (1977) studied 45 subjects with TS and their sisters. Twenty-four of the Turner subjects were 20 years old or more; and although these women had educational and occupational achievement comparable to their sisters', they were less likely to have moved out of their parental home, to have married, or to have established what the investigators considered to be "good sexual relations."

The study we report is the first to assess psychological functioning in a group of adult women with TS compared to a group of women with constitutional short stature (CSS). Women with CSS were chosen for comparison to control for the effects of growing up with visible short stature and of receiving a medical evaluation that identifies one as a patient. These experiences may have sequelae in themselves that are not unique to TS and should not be attributed to it. Our goal was to characterize psychological problems and assets particular to women with TS.

METHODS

Subject Selection

This investigation was approved by the relevant Institutional Review Boards at Columbia-Presbyterian Medical Center. All participating subjects gave written informed consent.

Subjects with TS were selected from the practice records of two of the authors (A.M. and J.B.) and the Pediatric Endocrinology Clinic records of all patients between the ages of 18 and 40. Although some were still under treatment by physicians at the same medical center, others had been seen briefly for diagnosis only and had not been under treatment there for 10 or more years. All patients meeting the age criteria were located. When the non-English-speaking (1), severely mentally retarded (1), and geographically distant (6) patients were eliminated, 29 remained. Of these, 6 declined to participate, and 23 took part in the study.

We wanted to compare the Turner group to a control group of women who were closely matched in race, age, and socioeconomic status (SES), who were short as adults, and who as children had an endocrinological evaluation for short stature at the same medical center. This control group was to have undergone normal spontaneous puberty and to have had no need for medical treatment to induce growth. We selected women who had been referred as children for evaluation of short stature in whom no organic cause for their short stature could be found. These patients had been given the diagnosis of constitutional short stature (CSS).

Each CSS woman was contacted by letter and later by telephone. Women who were willing to participate and who matched with Turner women for age (within 5 years), SES (within 15 points on the Hollingshead [1975] Four-Factor Index), and race were included if their adult height was less than or equal to 156.2 cm (61½″). This height criterion is 7.5 cm below average adult female height in the United States, according to the National Center for Health Statistics (1976). The criterion was adopted because enough women with an adult height of 61½″ or less who had CSS diagnosed as children were available to match with the 23 TS subjects.

Initially 64 CSS women potentially meeting the matching criteria for age, SES, and race were selected and attempts were made to contact them. Of these, 25 have not yet been located, 2 lived at too great a distance, 5 were too tall as adults, and 7 had developed a debilitating chronic disease or central nervous system disease that could affect psychological functioning. Of the remaining CSS patients (25), 21 agreed to participate, 2 are still undecided, and 2 refused. The present report concerns the first 9 CSS women from this source who completed the research protocol. Data collection is continuing on the remainder of the sample.

Assessment Methods and Procedures

Evaluation included a variety of psychiatric and psychological assessment methods.Requiring about 8 hours, the evaluation usually was completed in 1 day, although a second session was sometimes scheduled. Psychopathology and psychological functioning were evaluated by the Schedule for Affective Disorders and Schizophrenia-Lifetime Version (SADS–L) (Spitzer & Endicott, 1979), a semi-structured interview lasting about 90 minutes. This technique generates a life-time history of diagnosable mental disorders by Research Diagnostic Criteria (Spitzer, Endicott, & Robins, 1978) and also elicits information about personality style and level of psychological and social functioning in the 5-year and 1-month periods prior to the evaluation.

Information gained from the SADS–L as well as from other parts of the evaluation was used to make another kind of assessment of psychological functioning, using the Global Assessment Scale or GAS (Spitzer, Gibbon, & Endicott, 1978). The examiner assigns the subject a rating on a spectrum from 1 to 100, summarizing a person's current psychological health for the last week or other specified period, as measured by everyday functioning and degree of psychiatric symptomatology. The period of time assessed is used as an indicator of the subject's overall psychological functioning. In this study, the month prior to evaluation was rated. Higher GAS scores, generally those over 81, also indicate traits usually associated with positive mental health such as warmth, social effectiveness, and integrity.

All SADS–L interviews were conducted by the first author (J.D.). The interviews were audiotaped and diagnoses and GAS scores independently co-rated by one of the authors (R.G.), with disagreements resolved by discussion.

Response to medical syndrome was evaluated with the Syndrome-Specific Interview for Turner Syndrome and Constitutional Short Stature (Downey & Ehrhardt, 1981), a 45-minute semi-structured interview eliciting information about the subject's understanding of her medical condition and her emotional response to it. Sections of the interview focus on the subject's current understanding of why she has TS or CSS, her memories of the time of her evaluation at the medical center, her experience with short stature and with delayed puberty and infertility, the responses of others to her situation, and current concerns. All interviews were performed by one investigator (J.D.), and were audiotaped to enable co-rating for consistency.

Data analysis was based on statistical case-control comparisons. Statistical analysis of data on age, SES, IQ, height, weight, and GAS scores was done using the two-tailed t test for matched pairs. To assess differences in psychiatric diagnoses and responses to syndrome, the binomial version of the McNemar chi^2 was used, as recommended by Siegel (1956) for cases where cell frequency is low. Because of the small sample size and the possibility that tendencies evident now may become statistically significant when the whole sample has been studied, we report all p's < .20.

RESULTS

Total Turner Group (N = 23)

The sample characteristics are shown in Tables 10.1, 10.2, and 10.3. The Turner women averaged 27.1 years in age. Social status, as reflected in the Hollingshead Four-Factor Index, was similar for the subjects and their parents. The index is based on education and occupation levels and takes marital status into consideration. The majority of the women were of middle-class socioeconomic background. Mean full-scale IQ for the 22 women receiving the WAIS–R was in the average range with mean Verbal IQ 13.4 points higher than Performance IQ. Seventeen of the TS women (74%) had never been married.

TABLE 10.1
Sample Characteristics of Total Turner Group (N = 23)

	N	Mean	SD	Range
Age at Interview (Years)	23	27.1	5.2	19.6–38.1
Subjects' Hollingshead 4F Index	18[a]	42.3	11.1	25.0–66.0
Parents' Hollingshead 4F Index	23	42.8	14.5	17.5–63.0
Wechsler Adult Intelligence Scale-Revised—IQ				
Verbal	22[b]	99.7	11.4	78–122
Performance	22	86.3	10.1	70–105
Full Scale	22	92.9	9.6	77–112
Current Marital Status	N	%		
Married	5	22		
Divorced	1	4		
Never Married	17	74		

[a]The remaining subjects were still students (4) or unemployed (1) and unmarried at the time of the interview.

[b]One Turner woman was tested with the WAIS (Wechsler, 1955) before the WAIS-R (Wechsler, 1981) became available: Full IQ 106, Verbal IQ 106, Performance IQ 105.

Mean height for the sample, measured 3 times for each subject on the Harpenden stadiometer and averaged, was 142.2 cm (55½ in.). Mean weight for 22 subjects was 48.3 Kg (106½ lb).

All subjects had karyotypes performed at our medical center. Thirteen (57%) were 45,X; 7 (30%) were 45,X/46,XX or other mosaics, and 3 (13%) were 46,XX with structural abnormalities in one X.

TABLE 10.2
Height, Weight, and Chromosomes of Total Turner Group (N = 23)

	N	Mean	SD	Range
Height (cm)	23	142.2	6.0	132.2–157.0
Weight (kg)	22[a]	48.3	9.2	25.8–74.5
Karyotype	N	%		
45,X	13[b]	57		
45,X/46XX	6	26		
45,X/46,XX/47,XXX	1	4		
46,XXp-	1	4		
46,XXqi	2	9		

[a]One subject's weight was not measured. She reported a weight of 72.5 kg.

[b]One subject had a karyotype of 45,X but large Barr bodies present in the sex chromatin on her buccal smear suggested that she might be a mosaic-45,X/46,XXqi.

TABLE 10.3
Hormone Replacement Therapy of Total Turner Group (*N* = 23)

| | Ever | | Time of Evaluation | |
	N	%	N	%
Cyclic Estrogen/Progesterone Therapy	21[a]	91	17[b]	74
Monthly Withdrawal Bleeding	21	91	17	74
Secondary Sex Characteristics, Reported (Breasts & Pubic Hair Both Tanner Stage IV or V)	—	—	20[c]	87

[a]Two women had never received cyclic hormone replacement—1 had refused it, 1 had dropped out of treatment before beginning it.

[b]The 6 women who were not receiving hormone replacement at time of evaluation included 2 women never treated, 2 who stopped therapy after developing hypertension, 1 who stopped after removal of an endometrial polyp, and 1 who stopped after indicating little interest in withdrawal bleeding.

[c]The 3 women lacking secondary sex characteristics included the 2 women never receiving cyclic hormone therapy, and 1 who had discontinued it after approximately 4 years.

Twenty-one of the TS women had experienced puberty, 20 after estrogen replacement therapy and 1 spontaneously. The latter subject subsequently required hormone replacement to maintain cyclic bleeding. At the time of the study 17 TS women were maintained on cyclic estrogen-progesterone therapy.

Secondary sex characteristics at time of the research evaluation were reported by the subjects using self Tanner ratings of degree of maturation in comparison to photographs, according to the method of Duke, Litt, and Gross (1980). Twenty women reported both breast and pubic hair development to be Tanner stage IV or V, consistent with an adult female appearance.

The data on psychiatric diagnosis are listed in Table 10.4. Seven women (30%) had at least one diagnosable mental disorder present at the time of the research evaluation. Diagnoses ongoing at the time of the evaluation are designated in the table as "current." Four women had major depressions, lasting 1 week or more.[1] One woman had a phobic disorder. This is listed under "Anxiety" because by *DSM III* standards (American Psychiatric Association, 1980), it is one of the anxiety disorders. Three women had a total of five "other psychiatric disorder" diagnoses—one had schizoid personality and schizotypal features, one had passive-aggressive personality, and one had anorexia nervosa and schizotypal features.

The table also shows *lifetime* prevalence of psychiatric disorder, a term

---

[1]By RDC criteria, major depressions of 1–2 weeks are classified as "probable" and major depressions of more than 2 weeks are "definite." We have grouped probable and definite episodes of major depressive disorder together.

TABLE 10.4
Psychiatric Diagnoses of Total Turner Group ($N = 23$)
Measured by Schedule for Affective Disorder and Schizophrenia—
Life-time Version (SADS-L)

	Current Diagnoses		Lifetime Diagnoses	
	N	%	N	%
Number of subjects with ...				
Any Diagnosis	7[c]	30	16[d]	70
No Diagnosis	16	70	7	30
Number of Subjects with any				
diagnosis of ...				
Depression—Major	4	17	11	48
Depression—Minor or Intermittent	0	0	4	17
Hypomania	0	0	0	0
Anxiety[a]	1	4	5	22
Alcoholism	0	0	0	0
Drug-Use Disorder	0	0	1	4
Other Psychiatric Disorder[b]	5	22	6	26

[a]Includes panic disorder, phobias, generalized anxiety disorder, and obsessive-compulsive disorder.

[b]Includes personality disorders, anorexia nervosa, schizotypal features, and atypical depressive disorder.

[c]Subject is counted once if she had at least one diagnosable psychiatric disorder at time of evaluation.

[d]Subject is counted once if she had at least one diagnosable psychiatric disorder in her lifetime.

indicating the proportion of women who had ever experienced that disorder up to the date of evaluation. On the table a woman is counted once if she ever received a diagnosis, but recurrent episodes of the same disorder are not counted. Sixteen of the TS women (70%) had had at least one diagnosable mental disorder in their lifetime. Eleven of these had had at least one major depression, lasting 1 week or longer. The other diagnoses were largely minor or intermittent depressive disorders, anxiety disorders, or "other psychiatric disorders."

For the month prior to evaluation, psychological functioning as rated by the Global Assessment Scale or GAS for the 23 Turner women is shown in Table 10.5. Three subjects had scores less than 60, a score denoting major impairment. Seven subjects received GAS ratings of 81 or above, representing good to superior functioning. The majority of subjects fell in the middle range between 61 and 80, and the mean GAS score was 69.9.

TABLE 10.5
Current Psychological Functioning of Total Turner Group (N = 23)
Measured by the Global Assessment Scale (GAS)

		N	%
0-60	Moderate to major impairment and symptoms. Obvious to associates.	3	13
61-70	Some symptoms or difficulty functioning.	12	52
71-80	Slight impairment, minimal symptoms.	1	4
81-100	Good-superior functioning. Symptoms absent or transient. Satisfied with life.	7	30
GAS MEAN:			69.9

Turner (TS) Versus Constitutional Short Stature (CSS) Subjects (9 Matched Pairs)

The sample characteristics of the 9 matched pairs are listed in Tables 10.6 and 10.7. TS women were 1 year older on the average than their CSS controls. TS and CSS women were not significantly different with regard to their own or their parents' socioeconomic status. Seven of the TS women and eight of the CSS women had never been married; this was not a significant difference.

Although TS women were significantly lower in Performance IQ on the WAIS-R, Verbal IQ and Full-Scale IQ did not differ significantly.

Despite efforts to obtain CSS subjects who were as short as possible, the CSS women were 8.2 cm taller than the TS women. Mean weights were roughly comparable although the TS women had a much larger range. One CSS woman refused to allow measurement of height and weight, and a second was pregnant. Reported values for these subjects, using the pre-pregnancy weight for the pregnant subject, were used in calculating mean height and weight.

Table 10.8 shows the data on psychiatric diagnosis. Because of the small number of subjects and the low frequency of specific diagnoses, statistical tests were done only for the presence or absence of any diagnosis. At the time of the research evaluation, two of the TS women (22%) as compared to five of the CSS women (56%) had a diagnosable mental disorder. This was not a significant difference.

Six of the TS women (67%) and seven of the CSS women (78%) had at least one diagnosable mental disorder over their lifetime. Although these numbers are not significantly different, CSS women had twice as many diagnoses (10 vs. 20) with more non-major depressive disorders, anxiety disorders, and episodes of alcohol and drug abuse than the TS women. On the other hand, two TS women had long-standing severely impairing "other

TABLE 10.6
Sample Characteristics of Nine Matched Pairs:
Subjects with Turner Syndrome and Constitutional Short Stature (CSS)

	Turner (N = 9)				CSS (N = 9)				t test
	N	Mean	SD	Range	N	Mean	SD	Range	p (2-tailed)
Age at Interview (Years)	9	26.0	5.5	19.6–35.6	9	25.0	4.3	19.0–31.9	.09
Subjects' Hollingshead 4F Index	5[a]	46.7	10.1	31.0–56.0	6	45.1	10.9	28.0–56.0	NS[b]
Parents' Hollingshead 4F Index	9	48.5	14.3	17.5–63.0	9	54.4	7.5	41.5–66.0	NS
Wechsler Adult Intelligence Scale-Revised — IQ									
Verbal	9	106.6	12.0	82–122	9	109.2	11.6	91–131	NS
Performance	9	90.8	10.8	72–105	9	102.3	10.1	87–116	.07
Full Scale	9	99.3	10.2	77–112	9	106.6	10.9	89–122	NS

	N	%			N	%			Binominal Chi^2 p (2-tailed)
Current Marital Status									
Married	2	22			1	11			NS
Never Married	7	78			8	89			

[a]The remaining subjects were still students and unmarried at the time of the interview.

[b]NS = $p \geq .20$

TABLE 10.7
Height and Weight of 9 Matched Pairs:
Subjects with Turner Syndrome and Constitutional Short Stature (CSS)

| | Turner (N = 9) | | | | CSS (N = 9) | | | | t test |
	N	Mean	SD	Range	N	Mean	SD	Range	p (2-tailed)
Height (cm)	9	141.9	6.2	132.2–149.2	9[a]	149.1	5.3	141.7–157.7	.001
Weight (kg)	9[a]	49.3	12.8	25.8–72.5	9[a]	49.6	6.5	36.4–57.8	NS[b]

[a]One Turner subject's weight and 1 CSS subject's height and weight were not measured. Reported values are used for these subjects.
[b]NS = $p \geq .20$

TABLE 10.8
Psychiatric Diagnoses of 9 Matched Pairs:
Subjects with Turner Syndrome and
Constitutional Short Stature (CSS)
Measured by Schedule for Affective Disorders and Schizophrenia—
Life-time Version (SADS-L)

	Current Diagnoses				Lifetime Diagnoses			
	Turner		CSS		Turner		CSS	
	N	%	N	%	N	%	N	%
Number of subjects with . . .								
Any Diagnosis	2[c]	22	5	56[d]	6[e]	67	7	78[f]
No Diagnosis	7	78	4	44	3	33	2	22
Number of Subjects with any diagnosis of								
Depression—Major	1	11	0	0	4	44	4	44
Depression—Minor or Intermittent	0	0	4	44	1	11	5	56
Hypomania	0	0	0	0	0	0	1	11
Anxiety [a]	0	0	0	0	1	11	3	33
Alcoholism	0	0	1	11	0	0	1	11
Drug Use Disorder	0	0	1	11	0	0	4	44
Other Psychiatric Disorder[b]	3	33	2	22	4	44	2	22

[a]Includes panic disorder, phobias, generalized anxiety disorder, and obsessive-compulsive disorder.

[b]Includes personality disorders, anorexia nervosa, schizotypal features, and atypical depressive disorder.

[c]Subject is counted once if she had at least one diagnosable psychiatric disorder at time of evaluation.

[d]NS by binomial chi^2.

[e]Subject is counted once if she had at least one diagnosable psychiatric disorder in her lifetime.

[f]NS by binomial chi^2.

psychiatric disorders" (anorexia nervosa, passive-aggressive personality disorder), but no CSS women with "other psychiatric disorder" had a disorder that caused severe impairment (both CSS women had "other psychiatric disorder" because of the presence of schizotypal features, a diagnosis given them because of their belief in clairvoyance and extrasensory perception).

Table 10.9 shows Global Assessment Scale or GAS ratings for the 9 pairs. Although TS women tended toward the lower end of spectrum and CSS women toward the upper, mean GAS scores for TS women were 4.9 points lower, not a significant difference.

In the Syndrome-Specific Interview, subjects were asked how being

TABLE 10.9
Current Psychological Functioning of 9 Matched Pairs:
Subjects with Turner Syndrome and
Constitutional Short Stature (CSS)
Measured by the Global Assessment Scale (GAS)

		Turner (N = 9)		CSS (N = 9)	
		N	%	N	%
0–60	Moderate to major impairment and symptoms. Obvious to associates.	1	11	0	0
61–70	Some symptoms or difficulty functioning.	4	44	2	22
71–80	Slight impairment, minimal symptoms.	1	11	4	44
81–100	Good-superior functioning. Symptoms absent or transient. Satisfied with life.	3	33	3	33
GAS MEAN:			71.9		76.8[a]

[a]NS by 2-tailed t-test

short had affected their daily lives, with specific probes for inconvenience, embarrassment, and fright. As shown in Table 10.10, TS women tended to report inconvenience and fright less often than CSS women. Three TS and two CSS women reported embarrassment. However, two other CSS women, when queried about embarrassment, spontaneously answered, "No, but I get *angry.*" Two TS women reported that short stature had no effect on daily life. No such reports were heard from CSS women.

The Syndrome-Specific Interview also concerns the areas of the subject's life that have been most affected by her medical condition, whether TS or CSS (Table 10.11). Although subjects were asked to rank order areas of life affected (relationship with family of origin, career plans, relationship with

TABLE 10.10
Effect of Short Stature on Daily Life for 9 Matched Pairs:
Subjects with Turner Syndrome and Constitutional Short Stature
(CSS)

	Turner (N = 9)		CSS (N = 9)		Binomial chi² p (2-tailed)
	N	%	N	%	
Denies Effect	2	22	0	0	⎫ NS
Reports Effect[a]	7	78	9	100	⎭
Inconvenience	6	67	9	100	NS
Embarrassment	3	33	2	22	NS
Fright	2	22	4	44	NS

[a]Subjects falling into this category could rate more than one specific effect.

peers, and relationship with sexual partners) from most to least, not all could. However, all were able to pick the area of life most affected or to state that no area had been affected. CSS women were more likely to name relationship with sexual partners as most affected. Two TS women stated that no area of life had been affected at all. As indicated earlier, this was not the case with CSS women.

In summary, TS subjects had fewer psychiatric diagnoses, both current and lifelong, but lower psychological adjustment scores. They also tended to report fewer effects on their lives from their medical syndrome. All of these differences are tendencies only and do not reach statistical significance.

DISCUSSION

On the basis of Nielsen et al.'s and Taipale's reports on adolescent and adult women with Turner syndrome and on our own clinical impressions, we had expected TS women to have more psychopathology and more impairment in day-to-day functioning than CSS women. The lifetime prevalence of psychiatric disorder by SADS-L in the total TS group of 23 women shows that 44% have had at least one major depressive episode. Although this may seem high in view of the young age of the subjects, it is in line with the high prevalence of depressive disorders in young women reported by a number of major studies recently. For instance, using the same criteria for depression, the NIMH Collaborative Program on the Psychobiology of Depression found that 51% of female relatives of depressed patients had a depression by age 26–30 (Endicott, personal communication, March, 1984).

TS women compared to the CSS women had equal numbers of episodes of major depressive disorder. The number of TS and CSS women who had at least one diagnosable psychiatric disorder over their lifetime was almost comparable. However, CSS women had many more total diagnoses (20)

TABLE 10.11
Area of Life Most Affected by Syndrome for 9 Matched Pairs:
Subjects with Turner Syndrome and
Constitutional Short Stature (CSS)

	Turner (N = 9)		CSS (N = 9)	
	N	%	N	%
Relationship with Family of Origin	1	11	2	22
Career	3	33	1	11
Relationship with Peers	1	11	1	11
Relationship with Sexual Partners	2	22	5	56
No Effect	2	22	0	0

compared to the TS women (10). CSS women had more non-major depressive disorders, anxiety disorders, and episodes of alcohol and drug abuse. On the other hand, as already noted, two of the three TS women who had a diagnosis of "other psychiatric disorder" had chronic severe psychiatric impairment (anorexia nervosa, passive-aggressive personality disorder) but were unaware of it and complained of no subjective distress related to it.

The similar GAS ratings between TS and CSS women also suggest that the Turner group is not markedly more psychologically impaired. There is, however, a similar problem with interpreting the GAS scores as there is with the SADS–L data for the subjects with psychiatric disorders that do not produce subjective distress. The GAS score derives in part from a subject's degree of subjective psychiatric symptomatology. Subjects with no complaints tend to achieve better scores than those who report a great deal of subjective distress. If TS women report less subjective distress than do CSS women, their GAS scores may be artificially elevated.

These are preliminary results on psychiatric disorder in TS and CSS women because not all subjects scheduled for study have been evaluated. So far, findings on the nine pairs studied, if representative, suggest that TS women may have fewer diagnosable mental disorders. If they have suffered from mental disorders, they may be less likely to be of the "acting out" variety (e.g., alcohol and drug abuse) and more likely to be of the avoidant and withdrawn type (e.g., major depression with psychomotor retardation, anhedonia, and social withdrawal but without suicidal or other dramatic behavioral manifestations). A small subset of TS women may be severely impaired without necessarily being aware of it.

The response-to-syndrome data showed a trend that may be related to the psychiatric data already discussed. Although the TS and CSS women were well-matched for age, social class, and IQ, the TS women were more than 3 in. shorter on the average than the CSS women. Despite this, the TS women complained less about the effect of short stature on daily life. More CSS women than TS women noted that their relationships with sexual partners was the area of life most affected, a remarkable fact in itself since it is TS women, not those with CSS, who have fertility problems. This may be because some TS women admit to little interest in sexual relationships. The study also yielded data, not presented in this paper, demonstrating that TS women tend to be delayed in achieving psychosexual milestones such as dating and marriage. Thus, they may lack experience with opposite sex partners without reporting it or even perceiving it as such.

If the full study, when completed, confirms these early findings, we must then search for possible explanations. Several possibilities seem particularly likely: TS women may tend to minimize problems of which they are more or less aware. If so, it will be important to understand why they might do this. Second, as Money's term *inertia of emotional arousal* suggests, they may be protected in some way from feeling stress. The question then becomes whether the mechanism for this is an unconscious defense such as denial, or whether the stress is genuinely not experienced as stress. Related to these possibilities is the one that TS women may *avoid* potential experiences that could induce psychological stress. For instance, the suspicion that competing for male partners in a social setting might lead to rejection could lead TS women to avoid situations where they might encounter competition from other women for the attention of men.

Other possibilities for the lack of profound effects in TS women have to do with contacts with the medical care system. TS women may be protected from difficulties to some extent by the fact that they are under ongoing medical care. They are given a name for their condition, a reason for their short stature, and a treatment that is expected to be helpful. Having a problem such as Turner syndrome, which elicits medical interest and assistance, may be more conducive to optimism and good psychological adjustment than having a condition that is poorly understood or for which there is no treatment, as is the case in CSS. Finally, our subjects with CSS may be unusual in some way. For instance, CSS children with mild abnormalities of growth may be more likely to be referred for medical evaluation if their families are more disturbed or if they have had other family illness or problems that have led to medical referral in the past.

ACKNOWLEDGMENTS

This research was supported in part by BRSG GRANT #903-E582P of the Research Foundation for Mental Health, Inc., of New York State, and by NIMH Research Career Development Award #MH 00434 and Research Grants MH-34635 and MH-30906. Patricia Cohen, Jean Endicott, and Heino Meyer-Bahlburg were consultants on the project from its inception. Elizabeth Pierce, Patricia Rea, Roberta Stiel, and Matthew Tarran served as research assistants. Laura Rosen participated in the co-ratings of the interviews. Judith Feldman and Mel Janal were statistical consultants. Patricia Connolly and Dorothy Lewis provided secretarial services. We gratefully acknowledge the contribution of each of the above as well as the efforts and cooperation of the women who participated in the study.

REFERENCES

American Psychiatric Association (1980).*Diagnostic and statistical manual of psychiatric disorders, 3rd ed.* Washington, DC: Author.
Downey, J., & Ehrhardt, A. A. (1981). *Syndrome-specific interview for Turner syndrome and constitutional short stature.* Unpublished interview.

Duke, P. M., Litt, I. F., & Gross, R. T. (1980). Adolescents' self-assessment of sexual maturation. *Pediatrics, 66*(6), 918–920.

Ehrhardt, A. A., Greenberg, N., & Money, J. (1970). Female gender identity and absence of fetal gonadal hormones: Turner syndrome. *Johns Hopkins Medical Journal, 126,* 237–248.

Hollingshead, A. B. (1975). *Four Factor Index of Social Status.* Working paper. New Haven: Yale University.

Money, J. (1973). Turner's syndrome and parietal lobe functions. *Cortex, 9,* 385–393.

Money, J., & Mittenthal, S. (1970). Lack of personality pathology in Turner's syndrome: Relation to cytogenetics, hormones, and physique. *Behavioral Genetics, 1,* 43–56.

National Center for Health Statistics. (1976, June 22). *NCHS Growth Charts.* (Monthly Vital Statistics Report, 25 (3)). Rockville, MD: Health Resources Administration.

Nielsen, J., Nyborg, H., & Dahl, G. (1977). Turner's Syndrome: A psychiatric-psychological study of 45 women with Turner's Syndrome. *Acta Jutlandica, XLV,* 1–190.

Siegel, S. (1956). *Nonparametric statistics for the behavioral sciences.* New York: McGraw-Hill.

Spitzer, R., & Endicott, J. (1979). *Schedule for affective disorders and schizophrenia—Lifetime version.* New York: New York State Psychiatric Institute, Biometrics Research.

Spitzer, R. L., Endicott, J., and Robins, E. (1978, update 1980). *Research diagnostic criteria (RDC) for a selected group of functional disorders.* New York: New York State Psychiatric Institute, Biometrics Research.

Spitzer, R. L., Gibbon, M., & Endicott, J. (1978). *Global Assessment Scale* (GAS). New York: New York State Psychiatric Institute, Biometrics Research.

Taipale, V. (1979). *Adolescence in Turner Syndrome* (Private Publication). Helsinki, Finland: Children's Hospital, University of Helsinki.

Waber, D. P. (1979). Neuropsychological aspects of Turner's syndrome. *Developmental Medicine and Child Neurology, 21,* 58–70.

Wechsler, D. (1955). *Wechsler Adult Intelligence Scale.* New York: Psychological Corporation.

Wechsler, D. (1981). *Wechsler Adult Intelligence Scale—Revised.* New York: Psychological Corporation.

11 The Effects of Growth on Intellectual Function in Children and Adolescents

Darrell M. Wilson

Paula M. Duncan

Sanford M. Dornbusch

Philip L. Ritter

Ron G. Rosenfeld
Stanford University

The possible association between physical stature and intellectual development has fascinated investigators for decades. In his book, *Too Tall, Too Small*, Gillis (1982) lists a number of interesting anecdotes concerning the association between height and subsequent achievement. For example, all but two of the United States presidents since George Washington have been taller than the average height for their time and, since 1900, victory has gone to the taller presidential candidate in 80% of the electoral contests. Gowin (1916, pp. 22–33) demonstrated that within similar areas of employment, those in more prestigious positions were, on average, taller than those in positions of lesser prestige. Similarly, Boxer (1982) found a relationship between the heights of 5,085 Air Force cadets measured in 1943 and their annual income in 1968. In this report we utilize data from Cycle II and Cycle III of the National Health Examination Survey to determine if any associations exist between height and intellectual development and academic achievement among children and adolescents (Wilson et al., 1984). Additionally, using a longitudinal subset of these subjects, we have investigated whether growth rate has any significant association with changes in measures of intellectual development and academic achievement (Wilson et al., 1985).

METHODS

The National Health Examination Survey (NHES) was conducted during the 1960s by the National Center for Health Statistics to obtain health and psychosocial data from a national probability sample of the non-institutionalized United States residents (National Center for Health Statistics, 1974).During Cycle II of the NHES (1963–1965), 7,119 children, aged 6 through 11 years, were studied. Similarly, 6,768 adolescents, aged 12 through 17 years, were studied during Cycle III (1966–1970). In both samples, extensive data concerning the physical, intellectual, and psychosocial status of each subject were obtained. The samples were carefully stratified to represent the target population with respect to age, sex, race, region, population density, and population growth. By design, 2,177 eight-to eleven-year-old subjects first evaluated during Cycle II were reevaluated 2 to 5 years later during Cycle III, resulting in a large, well-selected longitudinal study population.

In order to compare heights among children and adolescents of varying age, we calculated a standardized index for height for each subject in Cycle II and Cycle III. This index, Z score for height, is the number of standard deviations that each subject's height differed from the mean height for that age (in months) and sex.

The degree of sexual maturation was determined for each subject in Cycle III by comparing the subject to a standardized set of photographs and assigning a Tanner stage for each female subject's breast and pubic hair development (Marshall & Tanner, 1969) and each male subject's genital and pubic hair development (Marshall & Tanner, 1970). These Tanner stages were then averaged to obtain a single sexual maturation score. Using this average score and the subject's sex and age, we divided the subjects into early, average, and late maturers (Duke et al., 1982). Early maturers were those whose sexual maturation score was greater than the 80th percentile for their cohort. Likewise, those subjects whose sexual maturation score was less than the 20th percentile were defined as late maturers. Those between the 20th and 80th percentile were defined as average maturers. For multiple linear regression models, the bone age/chronological age ratio was used as an index of relative maturation.

Socioeconomic status (SES) was estimated using family income. We defined less than $5,000 as low SES, $5,001 to $10,000 as mid SES, and greater than $10,000 as high SES. For multiple linear regression models, family income (dollars/year) was used directly as an index of SES. Each subject's racial background was characterized as white, black, or other.

The vocabulary and block design subtests of the Wechsler Intelligence Scale for Children (WISC) were used as a measure of intellectual ability and the reading and arithmetic subtests of the Wide Range Achievement

Test (WRAT) were used as a measure of academic achievement. These tests were administered during both Cycle II and Cycle III. Within the longitudinal subject sample, we calculated the change in normalized height, WISC, and WRAT scores by subtracting the values obtained during Cycle II from the values obtained during Cycle III.

Statistical analysis was performed using the Statistical Analysis System (SAS) Institute, Cary, North Carolina. Pearson's moment correlation coefficients were used to detect relationships between variables. Multiple linear regression models (with step-wise addition of variables) were used to determine the relative importance of height, sex, age, SES, race, and degree of sexual maturity on the WISC and WRAT scores. Because of occasional missing values, summation of the number of subjects within a given analysis may not equal the total number of subjects in that group. The p values reported have not been adjusted to compensate for multiple comparisons (Brown & Hollander, 1977, pp. 231–233).

RESULTS

Among the subjects in Cycle II, both WISC and WRAT scores showed a significant correlation with height ($r = .18$, WISC; $r = .17$, WRAT; p less than .0001 for each). Dividing the subjects into groups based on their sex, race, and sexual maturity did not affect the correlations between height and WISC and WRAT scores (see Table 11.1). Similarly, a significant correlation between age- and sex-normalized height and WISC or WRAT scores was seen among the adolescents in Cycle III ($r = .20$, $r = .19$, respectively, $p < .0001$ for each). This relationship is graphically illustrated in Fig. 11.1. IQ scores rise over 14 points from the shortest to the tallest group. As was observed in Cycle II, grouping by sex, race, and sexual maturation did not alter these correlation coefficients (see Table 11.1.). Interestingly, the association between height and IQ scores is less among the subjects with a high family income in both Cycle II and III.

We obtained a significant correlation between the weight Z score (weight normalized for sex and age) and IQ scores among the subjects in Cycle III ($r = .13$, WISC; $r = .12$, WRAT, $p < .0001$). However, when these correlations were adjusted for the correlation between the height and weight Z scores ($r = .59$) using the method of partial correlation (Sokal & Rohlf, 1981, pp. 656–661), it was clear that weight had little independent association with IQ score (adjusted $r = .01$ both for WISC and for WRAT).

To determine if the association between height and test scores for the subjects in Cycle III could be explained by the presence of illness in some of the subjects, we compared (using Student's t test) each of the following groups with the remainder of the sample: (1) whether the parents felt the

TABLE 11.1
Correlation Coefficients (*r*) Between
Height and WISC and WRAT Scores

	Cycle II		Cycle III	
	WISC	*WRAT*	*WISC*	*WRAT*
All	.18****	.17****	.20****	.19****
	(7,119)	(7,119)	(6,710)	(6,699)
By sex:				
Males	.19****	.18****	.20****	.20****
	(3,632)	(3,632)	(3,514)	(3,508)
Females	.17****	.16****	.20****	.18****
	(3,487)	(3,487)	(3,196)	(3,191)
By race:				
Whites	.17****	.17****	.22****	.21****
	(6,100)	(6,100)	(5,692)	(5,686)
Blacks	.16****	.13****	.18****	.15****
	(987)	(987)	(984)	(979)
By SES:				
Low	.19****	.20****	.17****	.18****
	(2,503)	(2,503)	(1,742)	(1,740)
Mid	.12****	.10****	.18****	.17****
	(3,103)	(3,103)	(2,624)	(2,619)
High	.07*	.09**	.12****	.08***
	(1,142)	(1,142)	(1,912)	(1,911)
By maturation:				
Early			.24****	.15
			(421)	(420)
Mid			.19****	.18****
			(5,279)	(5,271)
Late			.25****	.23***
			(787)	(785)

Note. Number of subjects is shown in parentheses.
*$p < .05$.   **$p < .01$.   ***$p < .001$.   ****$p < .0001$.   Hypotheses $r > 0$.

subject had had a severe illness ($n = 776$); (2) if the parents reported any lasting effects of this illness ($n = 325$); (3) if the birth had been abnormal ($n = 409$); (4) if the parents had any current health concern ($n = 986$); (5) if there had been any serious health concern since age one year ($n = 822$); (6) or whether the NHES medical examiner detected any current abnormalities ($N = 1,484$). There were no significant differences ($p > .05$) between any one of these subgroups and the remainder of the sample for the mean height Z score, WISC score, or WRAT score.

We further analyzed the relative contribution of SES (as family income in dollars/year), race, height Z score, age, and relative maturity (as the bone age/chronological age ratio) to IQ by utilizing multiple linear regres-

HEIGHT AND IQ

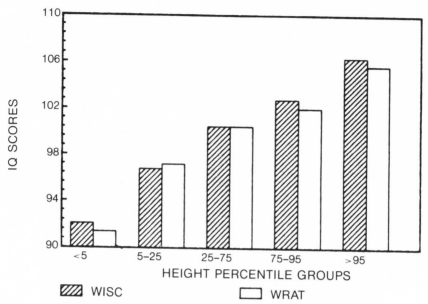

FIG. 11.1. Mean WISC and WRAT scores for the subjects in Cycle III divided into groups based on their height percentile.

sion with either WISC or WRAT as the dependent variable. The R squared values for males and females in both Cycle II and Cycle III are shown in Table 11.2. SES, race, and height were significantly associated with WISC and WRAT in Cycle II and III. Age did not contribute significantly to the model, and relative maturity was significant only among the females in Cycle III. In all cases, height contributed significantly ($p < .0001$) to the prediction of WISC and WRAT scores.

Figure 11.2 illustrates how the height Z score varied among the subjects within the longitudinal sample. The correlations between height and IQ scores in Cycle II and III, and the changes in these variables between Cycle II and III, are shown in Table 11.3. Although a correlation could be found between height and IQ scores in subjects from both Cycle II and Cycle III, no correlation could be detected between the change in relative height and change in either WISC or WRAT scores ($r = .03, r = .02$, respectively, $p > .2$ for each). Figures 11.3 and 11.4 illustrate this lack of association between the change in height and the change in IQ scores. Regrouping the subjects by sex, age, or relative height in Cycle II did not affect these results (see Table 11.4.). There is a slight correlation between the change in height and change in WISC score among the black subjects.

TABLE 11.2
Multiple Linear Regression Models (Cumulative R squared values)

	Males Cycle II		Cycle III	
	WISC	WRAT	WISC	WRAT
Family income	.220 ****	.172 ****	.217 ****	.205 ****
Race	.276 ****	.203 ****	.259 ****	.252 ****
Height Z score	.289 ****	.219 ****	.280 ****	.270 ****
Females				
Family income	.217 ****	.198 ****	.218 ****	.219 ****
Race	.263 ****	.223 ****	.262 ****	.274 ****
Height Z score	.275 ****	.233 ****	.283 ****	.293 ****
Bone age ratio	#	#	#	.295 *

Note.   Independent variables in this model: SES (as family income in dollars/year), race, height Z score, age, and relative maturity (as the bone age/chronological age ratio).
#–No other variables met the .05 significance level for entry into the model.
*$p < .05$.   **$p < .01$.   ***$p < .001$.   ****$p < .0001$.

DISCUSSION

A number of investigators have found an association between physical parameters and measures of intellectual development.Porter (1893) studied 33,500 students, aged 6 to 12 years, attending public school in St. Louis. At that time, students were promoted on the basis of school performance. Porter thus utilized grade placement as a measure of achievement and determined that heavier students did better than smaller students. Weinberg, Dietz, Penick, and McAllister (1974) found a correlation between head circumference and IQ ($r = .35$) in 334 eight-to nine-year-olds. Pollitt, Mueller, and Leibel (1982) demonstrated a significant correlation between IQ and the weight to height percentile in 91 three- to six-year-olds.

A number of other large studies have demonstrated significant correlations between IQ and height. The Scottish mental survey (Scottish Council for Research in Education, 1953, pp. 91–104) revealed a correlation between height and IQ ($r = .24$, males; $r = .2$, females) among 6,921 eleven-year-

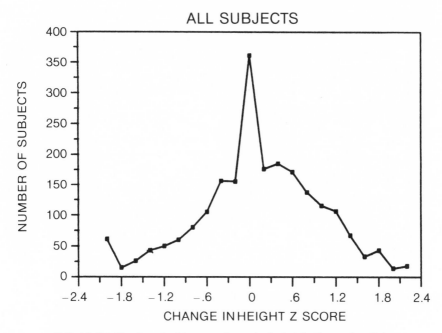

FIG. 11.2. Number of subjects vs. change in the height Z score for all the subjects in the longitudinal group (minus implies the subject became relatively shorter and plus implies the subject became relatively taller).

olds born in 1936. Miller (1974, pp. 223–228) found significant correlations between these same variables in a longitudinal study of 762 students born in 1947 and followed for 15 years. Douglas, Ross, and Simpson (1965) also observed a correlation coefficient of roughly .13 between height and IQ scores among 5,362 children born in Britain in 1946.

It is clear from these and similar studies that a multitude of other factors are also associated with IQ score. These include SES, race, relative maturity, and the presence of disease. The extensive data collected in Cycle II and Cycle III of the NHES have permitted us to examine the effects of these variables. In our study, we find a clear association between height and IQ scores that does not appear to be related to these other variables.

A possible explanation for the association between height and IQ scores is that both these parameters covary with the degree of physical maturation (as assessed by pubertal development or bone age). Thus, since physically advanced children tend to be taller at any given age, a correlation between relative physical maturation and IQ scores could then result in a correlation between height and IQ scores. In our study, however, no such effect of relative maturity could be detected. Moreover, as reviewed by Tanner (1969), significant correlations between height and IQ have also been

TABLE 11.3

Correlation Coefficients Between IQ Scores and Height in Cycle II and III and the Change in these Variables in the Longitudinal Group

	Cycle II			Cycle III			Change in Longitudinal		
	HGT2	WISC2	WRAT2	HGT3	WISC3	WRAT3	DHGT	DWISC	DWRAT
HGT2	1.00	.19 ****	.16 ****	.55 ****	.17 ****	.16 ****	− .45 ****	.03	.00
WISC2		1.00	.69 ****	.18 ****	.81 ****	.70 ****	.00	− .27 ****	.05 *
WRAT2			1.00	.16 ****	.68 ****	.83 ****	.00	.00	− .20 ****
HGT3				1.00	.17 ****	.16 ****	.45 ****	− .01	.02
WISC3					1.00	.74 ****	.02	.30 ****	.13 ****
WRAT3						1.00	.02	.06 **	.33 ****
DHGT							1.00	.03	.02
DWISC								1.00	.13 ****
DWART									1.00

$*p < .05.$ $**p < .01.$ $***p < .001.$ $****p < .0001.$ Hypotheses $r > 0.$

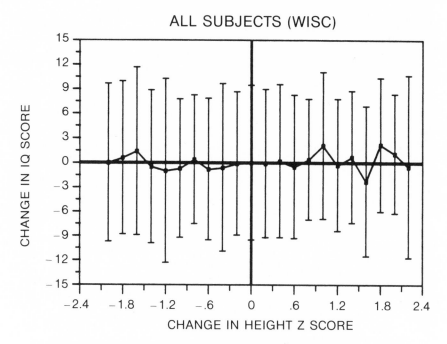

FIG. 11.3.   Change in WISC score (± 1 *SD*) vs. change in the height *Z* score for all the subjects in the longitudinal group.

observed in a number of studies of adults. Although these studies each have flaws, together they strongly suggest that the association between height and IQ found in our study persists into adulthood and thus does not reflect transient differences in relative maturity.

We found significant correlations between height and IQ in both Cycle II and Cycle III of the NHES. However, within the longitudinal subgroup, numbering over 2,000 subjects, no association between change in height and change in IQ score could be detected. Taken together, these data imply that the processes that contribute to the association between height and IQ must occur relatively early in childhood, or at least be established before the age of the children in the longitudinal sample.

The reasons for the association between height and IQ scores are unclear. It is possible that both these parameters covary with an undefined variable such as subtle intra-uterine or post-natal damage. On the other hand, the association between height and IQ scores may reflect a social response to relative stature of the child. It is possible that parents and others interact differently with short children than they do with taller children of the same age. The association between height and IQ score may result from the altered expectations of parents and adults based on the child's relative height for age.

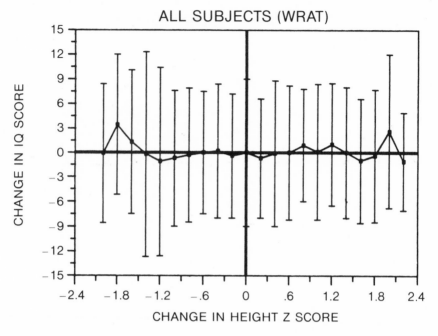

FIG. 11.4. Change in WRAT score ($\pm$ 1 *SD*) vs. change in the height *Z* score for all the subjects in the longitudinal group.

Further, this study demonstrates that naturally occurring changes in relative height between the ages of 8 and 13 years do not significantly affect IQ scores. Although the information obtained in this study cannot directly address the important question of whether or not medical therapy to change height will affect IQ scores, these data suggest that we should be cautious with our expectations of such treatment. Until carefully designed prospective clinical studies to evaluate these questions are designed and executed, any effect of growth hormone therapy upon intellectual development must be considered unlikely.

TABLE 11.4
Correlation Coefficients Between the Changes in Height
and Changes in IQ Scores

	Change in WISC	Change in WRAT
All	.03	.02
	(2177)	(2176)
By sex:		
Males	−.01	.03
	(1138)	(1137)
Females	.07*	.01
	(1039)	(1039)
By race:		
Whites	.00	.03
	(1897)	(1897)
Blacks	.17**	.03
	(271)	(270)
By SES:		
Low	.04	.01
	(632)	(631)
Mid	.03	.01
	(1033)	(1033)
High	−.03	.04
	(410)	(410)

Note.   Number of subjects is shown in parentheses.
*$p < .05$.   **$p < .01$.   Hypotheses $r > 0$.

REFERENCES

Boxer, A.(1982). In J. S. Gillis. *Too tall, too small.* Champaign: Institute for Personality and Ability Testing.

Brown, B. W., & Hollander, M. (1977). *Statistics.* New York: Wiley.

Douglas, J. W. B., Ross, J. M., & Simpson, H. R. (1965). The relationship between height and measured educational ability in school children of the same social class, family size, and stage of sexual development. *Human Biology, 37,* 178-186.

Duke, P. M., Carlsmith, J. M., Jennings, D., Martin, J. A., Dornbusch, S. M., Gross, R. T., & Siegel-Gorelick, B. (1982). Educational correlates of early and late sexual maturation. *Journal of Pediatrics, 100,* 633-637.

Gillis, J. S. (1982). *Too tall, too small.* Champaign: Institute for Personality and Ability Testing.

Gowin, E. B. (1916). *The executive and his control of men.* New York: Macmillan.

Marshall, W. A., & Tanner, J. M. (1969). Variations in the pattern of pubertal changes in girls. *Archives of Diseases in Childhood, 44,* 291-303.

Marshall, W. A., & Tanner, J. M. (1970). Variations in the pattern of pubertal changes in boys. *Archives of Diseases in Childhood, 45,* 13-23.

Miller, F. J. (1974). *The school years in Newcastle upon Tyne 1952-62: Being a further contribution to the study of a thousand families.* New York: Oxford University Press.

National Center for Health Statistics. (1974). *Plan and operation of a health examination survey of U.S. youths 12-17 years of age* (Vital and Health Statistics, Series 1, No. 8, pp. 1-80). Washington, DC: U. S. Government Printing Office.

Pollitt, E., Mueller, W., & Leibel, R. L. (1982). The relationship of growth to cognition in a well-nourished preschool population. *Child Development, 53,* 1157-1163.

Porter, W. T. (1893). The physical basis of precocity and dullness. *Transactions Academy Science, St. Louis, 6,* 161-181.

Scottish Council for Research in Education (1953). *Social implications of the 1947 Scottish mental survey.* London: University Press.

Sokal, R. R., & Rohlf, F. J. (1981). *Biometry.* San Francisco: W. H. Freeman.

Tanner, J. M. (1969). Relation of body size, intelligence test scores, and social circumstances. In P. H. Mussen (Ed.), *Trends and issues in developmental psychology* (pp. 183-193). New York: Holt.

Weinberg, W. A., Dietz, S. G., Penick, E. C., & McAllister, W. H. (1974). Intelligence, reading achievement, physical size and social class. *Journal of Pediatrics, 85,* 482-489.

Wilson, D. M., Duke, P. M., Dornbusch, S. M., Ritter, P. L., Carlsmith, J. M., Hintz, R. L., Gross, R. T., & Rosenfeld, R. G. (1984). Height and intellectual development. *Pediatric Research, 18,* 100A.

Wilson, D. M., Hammer, L. D., Duncan, P. M., Dornbusch, S. M., Ritter, P. L., Gross, R. T., Hintz, R. L., Rosenfeld, R. G. (1985). *Growth and intellectual development.* Submitted for publication.

12

Environmentally Based Failure to Thrive: Diagnostic Subtypes and Early Prognosis

Dennis Drotar
Case Western Reserve University School of Medicine

Environmentally based failure to thrive (FTT) is a frequent problem. Recent surveys report a 10%–20% prevalence in rural (Mitchell, Gorell, & Greenberg, 1980) and urban ambulatory care settings (Massachusetts Department of Public Health, 1983). In addition, FTT accounts for 1%–5% of pediatric hospitalizations of young children (Berwick, 1980; Hannaway, 1970; Homer & Ludwig, 1980; Sills, 1978). In contrast to conditions that disrupt children's psychological development and that cannot be identified early in life, the screening of physical growth via well-child visits provides an objective means to identify children at risk for chronic deficits in physical growth, cognitive and emotional development. Despite the limitations in follow-up studies owing to small sample sizes and retrospective methods (Drotar, Malone, & Negray, 1979), children with an early diagnosis of FTT have a greater than average chance of developing problems in physical and psychological functioning. Table 12.1 depicts risk domains in which the FTT child's functioning may be compromised and potential environmental risk factors (Drotar, in press) that may affect growth and psychological development.

Physical Growth

Children identified as failing to thrive have variable rates of physical growth following diagnosis and treatment.Some clinicians have reported rapid changes in rate of weight gain (Field, 1984) following hospitalization and aftercare, but others have reported little change after one-year follow-up

151

TABLE 12.1
Biopsychosocial Functioning and Risk in Failure to Thrive

	Child	
Physical Status	*Psychological Status*	
physical growth	cognitive development	
nutritional status	behavioral reaction to test objects	
health status	social behavior	
physical symptoms	affective development	
	attachment	
	behavioral symptoms	
	Environment	
Proximal Environment	*Family Environment*	
physical environment	income and resources	
level of stimulation	size	
child's social network	structure	
physical care	number and spacing of siblings	
frequency and structure of	parental emotional development	
parent-child interaction	intrafamilial relationships	
	functioning stresses	
	social network	

even with intervention (Fitch et al., 1975). Long-term physical growth outcomes in FTT range from 20% (Shaheen, Alexander, Truskowsky, & Barbero, 1968) to 60% of children (Elmer, Gregg, & Ellison, 1969) remaining below the 3rd percentile in height and/or weight on follow-up 3–5 years after hospitalization. However, the factors that affect long-term prognosis with respect to physical growth are not known.

Cognitive Development

Cognitive developmental deficits associated with FTT range from moderate delays (Leonard, Rhymes, & Solnit, 1966) to a high frequency of mental retardation (Ramey, Starr, Pallas, Whitten, & Reed, 1975). Although controlled studies are rare, most long-term studies show high levels of cognitive impairments among FTT children compared to test norms (Breunlin, Desai, Stone, & Swilley, 1983). Follow-up studies report varying outcomes for subsequent intellectual development. Field (1984) reported that average Bayley MDI improved from 78 to 97 at 3-month follow-up and was maintained on 6- to 13-month follow-up. On the other hand, very different initial and long-term findings were reported by Singer and Fagan (1984) who found that FTT children originally evaluated at 8 months of age maintained their below average intellectual levels (MDI below 80) at 20 and 36 months of age. Moreover, there was a significant difference between children who returned home following hospitalization (MDI of 90) versus those placed in foster care (MDI of 70).

Emotional Development

Controlled studies indicate a higher number of behavioral problems in FTT children compared with physically healthy preschoolers (Pollitt & Eichler, 1976) and diminished responsiveness to test objects and goal-direct behavior compared with abused infants (Fitch et al., 1975). Older children with histories of FTT demonstrate severe problems with behavioral control (Leonard et al., 1966; Money, Wolff, & Annecillo, 1972; Powell, Brasel, & Blizzard, 1967; Silver & Finkelstein, 1967). In addition, studies document a high rate of behavioral disturbances (27%–48%) among school-aged children (Elmer et al., 1969; Glaser, Heagarty, Bullard, & Pivchik, 1968; Hufton & Oates, 1977) as judged by clinicians or teachers. FTT children who have suffered the combined effects of chronic neglect and abuse are at special risk for behavioral disorders. Children with psychosocial dwarfism demonstrate such disorders as sleep disturbances, enuresis, encopresis and absence of appropriate pain response, self-harming behaviors, social distancing and temper tantrums (Money & Annecillo, 1976; Money, Annecillo, & Werwas, 1976; Money, Annecillo, & Kelley, 1983; Wolff & Money, 1973).

Attachment Behavior

Given the significance of a secure child-to-parent attachment for subsequent socialization, problem solving, and emotional competence (Matas, Arend, & Sroufe, 1978; Pastor, 1981; Sroufe, 1979, 1983; Waters, Wippman, & Sroufe, 1979) it is quite possible that some of the behavioral deficits found in preschool children with FTT (Pollitt & Eichler, 1976) may be mediated by compromised early attachments.Available evidence is consistent with clinical observations that suggest disruptions in attachment among FTT children. For example, using a paradigm similar to Ainsworth & Wittig (1969), Gordon & Jameson (1979) found that significantly more nonorganic FTT children were insecurely attached to their mothers than control children.

Individual Differences and Psychosocial Diagnosis

Although the high level of risk associated with an early diagnosis of environmentally based FTT is reasonably well established, the paucity of prospective studies of treatment outcome and prognosis in this condition represents a significant gap in scientific knowledge. It is reasonable to assume that not every child who is diagnosed with FTT has an equal probability of developing chronic deficits in psychological development or physical growth. In accord with an interactional model of development (Sameroff & Chandler, 1975), the probability of long-term risk associated with FTT should be a function of individual differences in physical status (especially duration and severity) *and* the quality of the child's family environment. For this reason, documentation of individual differences in physical status and the relationship to psychological development assumes importance.

Clinicians have begun to recognize that environmentally based FTT is not a unitary entity but a heterogeneous condition that encompasses a number of subtypes not heretofore recognized. Significant direction for future research involves classification of physical and psychological influences in FTT and the relationship to prognosis. The complex, psychosomatic nature of FTT has defied comprehensive classification of prognostic factors partly because of the profession-specific nature of diagnostic tasks. For example, pediatric diagnosis generally involves differentiation between environmental versus organic etiologies of FTT. On the other hand, psychiatric diagnosis, e.g., Reactive Attachment Disorder (RAD) (American Psychiatric Association, 1980) identifies deviant emotional development among children with environmentally based FTT. Major criteria of the diagnosis of RAD include deviant emotional development, especially lack of age-appropriate signs of social responsiveness, environmentally based growth deficit, and onset before 8 months. Unfortunately, the utility of the RAD diagnosis to describe psychological prognosis is limited by a lack of specificity, which makes it extremely difficult to reliably assess impaired social responsiveness. In addition, the age-specific nature of RAD, which is restricted to children who are 8 months and under, limits the generalizability of this diagnosis to describe the spectrum of environmental FTT, especially the large numbers of children who present to practitioners after the age of 8 months.

In recent years, a number of classification schema have been proposed to refine and extend the RAD diagnosis (Egan, Chatoor, & Rosen, 1980; Woolston, 1983). For example, Egan et al. (1980) proposed three major subtypes based on etiology: (1) underfeeding, characterized by a lack of appropriate nutrition in the absence of maternal psychopathology or problems in the maternal-child relationship; (2) deficiency in maternal care (reactive attachment disturbance); and (3) disturbance in separation and individuation, which is characterized by onset during the second half of the first year of life and severe feeding disturbance (Chatoor & Egan, 1983). Woolston's (1983) typology, which includes Type I (reactive attachment disorder), Type II (simple protein-calorie malnutrition), and Type III (pathological food refusal), also includes variables presumably relevant to etiology such as maternal psychopathology, maternal appraisal of the child's condition (e.g., whether the child is perceived as physically sick), and the quality of the child's social interaction. Although such etiology-based classifications suggest interesting hypotheses for empirical study of diagnostic subtypes, objective decision rules are necessary for their application by clinicians and researchers. In the absence of specific criteria, it is difficult to differentiate among presumed etiologies. The problem of establishing etiology in environmentally based FTT is especially difficult because clinical observations and history are gathered after the condition has become chronic and/or the child is separated from his or her family context in a hospital setting.

Given the difficulties of cause and effect determinations in children who have developed FTT, objective assessment of physical status and course, including such variables as age of onset and duration, should have greater potential generalizability than an etiology-based classification schema. Environmentally based FTT may have very different origins and consequences depending on age of onset (Chatoor & Egan, 1983; Woolston, 1983). For example, early onset FTT may represent a disturbance in early regulatory patterns between mother and child (Sander, 1964) whereas later onset FTT can reflect parental difficulties in responding to the child's initiatives for independence. Duration or chronicity may also affect the severity of the child's physical and/or psychological deficits at point of presentation and eventual response to treatment. In practice, there is often a significant time span between age at which the child's physical growth is first affected and the age at diagnosis. Because some FTT infants are lost to pediatric follow-up and the diagnosis of environmentally based FTT is difficult to make, diagnosis and treatment are often delayed relative to the onset of the condition. The discrepancy between onset and identification of FTT may have salient developmental consequences. For example, the longer the child's FTT progresses, chronic undernutrition limits opportunities for brain growth and stimulation (Pollitt, 1969). By the time the child's weight gain is sufficiently compromised to warrant pediatric hospitalization, malnutrition may be severe enough to lower social responsiveness, influence caregivers' response to the child, and potentially long-term psychological outcomes (Pollitt, 1969, 1973). The present report from an ongoing prospective study of outcome in environmentally based FTT describes the early outcome of children categorized according to age of onset and duration. The major hypothesis is that early onset FTT, which is also of relatively long duration, will be associated with greater psychological impairment at outcome than FTT that is identified shortly after onset.

METHOD

This report concerns early outcome (intake to 18 months) from an ongoing prospective study in which FTT children and a matched sample of physically healthy children are followed from point of intake, on the average 5 months of age, to age 3½.The study has the following goals: (1) early treatment and prospective follow-up of children with environmentally based FTT, (2) identification of factors (treatment and non-treatment) that are associated with physical and psychological outcomes in this condition, and (3) identification of FTT children at special risk for behavioral and developmental disorders.

Selection Criterion

In accord with recommendations for pediatric diagnosis (Bithoney & Rathbun, 1983; Schmitt, 1978), children were judged as FTT if they demonstrated (1) weight at or below the 5th percentile based on norms (Hamill et al., 1979); (2) absence of major organic conditions that could directly affect capacity to gain weight and/or cognitive development as indicated by physical and laboratory examination, including complete blood count and urinalysis; (3) demonstration of weight gain in hospital (Schmitt, 1978); and (4) decrease in rate of weight gain from within normal limits at birth to below the 5th percentile. Children who demonstrated slow but constant patterns of growth consistent with constitutional deficits were not included. Children with lower than average growth potential were excluded by requiring growth parameters (height, weight and head circumference) to be appropriate for gestational age at birth, and a birth weight of at least 1500 g. Three criteria increased sample homogeneity and simplified data collection: (1) age between 1 and 9 months of age; (2) absence of child abuse (three children were excluded for this reason); and (3) geographic proximity (within an hour's distance from Cleveland).

Subject Recruitment and Sample Attrition

Subjects were recruited from children hospitalized for FTT at one of seven Cleveland area hospitals.Seven families (out of 83 with children who fit the criteria) did not choose to participate and did not differ in demographic characteristics from the study group. Of the remaining 76 families, 11 (15%) dropped out because they moved out of the area, could no longer be located, or refused further participation. This included one study child judged to be in danger of abuse and needing custody of county welfare protective services. The attrition sample did not differ significantly from the study group in demographic characteristics, age, physical growth or cognitive development at study intake.

Subjects

The mean age (n = 68) at study intake was 4.9 months (range 1–9) for 44 males and 24 females from 65 families, including three sets of twins. Racial composition included 38 black, 26 white, and 3 children of Spanish heritage. Children tended to be latter borns (mean birth order = 2.6) from disadvantaged families in urban neighborhoods. Fifty-four families (82%) received welfare (Aid to Dependent Children). Other families (n = 11) were working class (M = 3.7) on Hollingshead and Redlich Scale (1958). Other family demographics were as follows: mean income, $5,800, family size ($M$ = 5), number of children per family (M = 2.5), ratio of adults to children (M = 1.3), and maternal educational level (M = 10.9) school years.

Intervention

Following hospitalization, families were randomly assigned to one of three interventions.All involved working with family members in their homes but differed with respect to frequency of contacts and focus. In one intervention plan, family members were seen for an average of six visits over a year. In two others, parents or family members were seen for weekly home visits for an average duration of 1 year as follows: Family-centered (Drotar & Malone, 1982) intervention involved relevant members of the family group to enhance family coping and support and modify dysfunctional patterns of intervention to enhance child nurturing. Parent-centered intervention involved a supportive educational focus directed toward improving the mother's (or major caretaker's) interactions, nutritional management, and parent-child relationship. Although family-centered intervention was expected to result in more permanent gains once treatment had ended, no differences were predicted on initial outcomes (12–18 months) which were evaluated while families still received intervention.

ASSESSMENT PROCEDURES

Outcome measures were chosen for reliability of scoring, feasibility of administration, and sensitivity to psychological deficits associated with FTT. Intake assessments were conducted in the hospital. Physically compromised children were not evaluated until they improved to the point where they could respond to the test items (Drotar, Malone, & Negray, 1980). Outcome assessments were conducted in family homes by experienced examiners who were unaware of treatment assignment or information about families.

Physical Growth

Infants were weighed on equivalent scales (Health-O-Meter Pediatric Scale, Continental Model No.322). Head circumference was assessed by the Inserta Tape, Ross Laboratories. All measurements were taken at study intake, 12 and 18 months. Two measures were calculated from weight, length, and head circumference. *Wasting* was percentage of the child's weight typical for a given height, as specified by growth charts. Four general classifications were used (Waterlow, 1972; Waterlow & Rutishauser, 1974): 0 = greater than 90% weight for height; 1 = 80%–89%; 2 = 70%–79%; 3 = less than 70%. *Stunting* refers to the percentage of expected length for age and was calculated by comparing the child's length with norms. The following classification schema was used (Waterlow, 1972): 0 = greater than 95% height for age; 1 = 90%–95%, 2 = 85%–89%; 3 = less than 85%.

Family Income and Structure

For families receiving welfare, income was determined from standard payments.Variations for the families on welfare corresponded to the number of eligible children and/or differences in social security payments. Family size and the ratio of adults to children living in a household were corroborated with multiple observers.

Intellectual Development

Cognitive development was assessed with the Bayley Scale of Mental Development (Bayley, 1969) at study intake and at 12 and 18 months.

Attachment

Attachment was assessed at 12 months of age via the Strange Situation, a laboratory procedure developed by Ainsworth and Wittig (1969) which consists of eight episodes presented in a standard order for all subjects in an 11 × 14 ft.room. The order of episodes is so arranged that the infant experiences a series of increasingly (mildly) stressful situations. Episodes are videotaped and the following dimensions rated: (Ainsworth, Blehar, Waters, & Wall (1978)

1. *Proximity seeking*—The intensity and persistence of the baby's efforts to gain (or regain) physical contact (or more weakly, proximity) with an adult.

2. *Contact maintaining*—The degree of activity and persistence in the baby's efforts to maintain physical contact with an adult once he has gained it (especially such active resistance to being released as clinging or protesting); also, behaviors such as sinking in while held, which tend to delay the adult's attempts to release the baby (i.e., to prolong contact by not signaling readiness for release).

3. *Proximity and interaction avoiding*—The intensity, persistence, duration, and promptness of any active avoidance of proximity or interaction, even across a distance, especially in reunion episodes. Included here are aborted approaches upon reunion, turning the face away when greeted, prolonged pout and refusal to make eye contact or to interact, and mild signs of wariness of the stranger accompanied by retreat to the mother.

4. *Contact resisting*—The intensity and frequency or duration of negative behavior evoked by a person who comes into contact or proximity with the baby, especially behavior accompanied by signs of anger. Relevant behaviors include pushing away, dropping or hitting toys offered, body movements in resistance to being held. This behavior may alternate with active efforts to achieve or maintain contact, and both can be scored high in the same episode.

These ratings are used to define a pattern of attachment as follows: *secure*—high-proximity seeking and maintaining with low avoidance or resistance—and *insecure*. There are two types of insecure attachment: (1) *avoidant*, low-proximity seeking and high avoidance, and (2) *resistant* (ambivalent), which includes high-proximity seeking and resistance. Security of attachment relates to quality of problem solving, competence with peers, and sociability during the preschool years (Matas et al., 1978; Pastor, 1981; Waters et al., 1979). Videotapes were judged by two raters who were blind as to the subject's physical status and attachment classification. The inter-rater percentage of agreement for the behavioral dimensions was 86%.

Behavioral Ratings

Assessors filled out rating scales adapted from the Infant Behavior Record (Bayley, 1969) which assess the child's behavior during testing on a 5-point scale (a score of 5 was the most adaptive rating) in the following areas: Cooperativeness, Object Orientation, Social Orientation, and Goal Directedness. Inter-rater reliability ranged from 85% to 95%. Correlations between individual subscales and total composite ranged from .74 to .89. A composite score was utilized in this analysis.

Language Ability

Language ability was determined in children 18 months and older via a battery used by White, Kohan, Antanucci, & Shapiro (1978) in a study of the competence of preschoolers from different socioeconomic backgrounds. Derived scoring categories, including vocabulary and comprehension of grammar, were shown to have adequate inter-rater reliability (.90 to .99). Inter-rater reliability was consistently above 90% in the present study.

Symbolic Play

Capacity to represent experience in the non-verbal medium of play was assessed by the Symbolic Play Test (Lowe, 1975), a structured procedure in which the child is given sets of play objects according to a standardized format and observed in spontaneous play.Standardized on a sample of 240 children, this test has acceptable internal consistency (.92) and test-retest reliability (.81) (Lowe & Costello, 1976). Scoring categories based on objective, discrete behaviors assess the complexity of the child's play from very simple usage such as picking up a toy to relating objects to one another. Inter-rater reliability was 95%.

RESULTS

Physical Growth and Development at Intake

The means and standard deviations for physical growth and psychological variables are shown in Table 12.2.

At hospital admission, mean rate of weight gain was more than two standard deviations (less than the 5th percentile) below the norms for age (Hammill et al., 1979) compared to a mean birth percentile of 32.7. Although physical growth ranged widely at study intake, 59 (87%) showed at least a mild degree of wasting ($M = 1.5$, $SD = 0.8$) and 32 (53%) a mild or greater degree of stunting ($M = 0.7$, $SD = 0.7$). On the other hand, intellectual functioning based on the Bayley Mental Development Index (MDI) ($M = 99.4$, $SD = 16.8$) was within normal limits.

Physical and Psychological Outcome

As shown in Table 12.2, mean scores on the Bayley (MDI), Behavioral Ratings, Symbolic Play Test and Language Ability at 12 and 18 months were within normal limits. In addition, the physical growth data indicate

TABLE 12.2
Physical Growth and Psychological Measures

	Intake (N = 68)		12 Months (N = 67)		18 Months (N = 65)	
	M	SD	M	SD	M	SD
Height (cm)	59.7 (20th %ile)	6.5	72.5 (25th %ile)	2.9	79.1 (26th %ile)	3.0
Weight (kg)	4.6 (5th %ile)	1.3	8.6 (21st %ile)	1.1	10.1 (27th %ile)	1.2
Head Circumference (cm)	39.9 (25th %ile)	3.2	45.7 (29th %ile)	1.5	47.2 (38th %ile)	1.3
Bayley MDI	99.3	16.3	109.6	15.1	102.4	13.6
Behavioral Ratings	11.2	2.6	13.8		12.1	3.0
Language	–	–	–	–	17.8	3.3
Symbolic Play	–	–	–	–	7.7	2.6

recovery with growth parameters in the normal range. To assess the effect of treatment modality on early outcome, multivariate analyses of variance (MANOVA) were conducted on physical growth (height, weight, and head circumference), Bayley MDI and Behavioral Ratings at intake, 12 and 18 months. One-way analyses of variance were also conducted for Symbolic play and Language Ability at 18 months. None of these analyses revealed main effects of type of treatment.

Attachment

The following attachment classifications were found at 12 months of age: B (Secure) 33 (49%), A (Avoidant) 22 (33%), C (Resistant) 8 (12%), and (Unclassified) 4 (6%).The same frequency of infants (49%) were securely attached as were insecurely attached (45%), which is a higher rate than the 20%–30% rate of insecure attachment reported for physically healthy populations from low risk family environments (Campos et al., 1983).

ANALYSIS OF INDIVIDUAL DIFFERENCES: DURATION AND SEVERITY

An estimate of age of onset, defined as the age at which the child's rate of weight gain reached the 5th percentile, was made via detailed analysis of the child's growth curve including birth and pediatric records.A majority of children had at least three points on the growth curve. This definition of onset is a conservative one because it underestimates the age at which deceleration in rate of weight gain occurs. By definition, children who reached the 5th percentile dropped off from a birth weight that was within normal limits. Duration was defined as the time that elapsed between the age at which the child reached the 5th percentile in weight and age at study intake (pediatric hospitalization). This analysis compared two groups: Early Onset and Chronic ($N = 17$) versus Late Onset, Non-Chronic ($N = 12$). The groups were defined empirically on the basis of a median split for onset and duration variables separately. The weight of the Early Onset group dropped off quite early relative to age expectations. Average age of onset was 1.2 months and duration ($M = 4$ months). The Late Onset group reached the 5th percentile later ($M = 6$ months). Average duration was of less than a month in the late onset group, which means that diagnosis and treatment were initiated quite early in the course of the condition. Table 12.3 shows the differences in the two groups on physical growth, and psychological development at study intake and 18 months of age, which are the latest completed available outcome data on the entire sample.

TABLE 12.3
Psychological and Physical Status at Intake and 18 Months

	Early Onset— Chronic			Late Onset— Nonchronic	
			Intake		
	Mean	SD		Mean	SD
Weight (kg)	4.2	0.9		5.71***	.8
Length (cm)	9.1	5.3		64.7***	4.1
Head Circumference	39.9	2.8		41.9*	1.3
Stunting	1.2***	0.8		0.2	0.3
Wasting	1.6	0.9		1.3	0.8
Bayley MDI	94.4	16.7		110.2*	19.9
Behavioral Ratings	10.9	2.7		11.2	3.1
			18 Months		
Weight (kg)	9.7	11.0		9.9	11.1
Length (cm)	7.8	3.6		7.8	1.9
Head Circumference (cm)	47.3	1.3		46.9	0.9
Stunting	0.6	0.7		0.3	0.4
Wasting	0.2	0.4		0.2	0.5
Bayley MDI	95.6**	13.2		110.0	14.3
Behavioral Ratings	11.8	2.5		12.5	3.3
Language Ability	17.1	3.5		17.6	1.6
Symbolic Play	7.1	1.8		9.1	3.3

* $p < .025$
** $p < .01$
***$p < .005$

As shown in Table 12.3 at point of study intake, the Early Onset Chronic group was more compromised in all parameters of physical growth and cognitive development than the Late Onset, Non-Chronic group. Of particular importance is the lower Bayley MDI (M = 94.4) in the Early Onset versus Late Onset group (M = 110.2, t = 2.28, df = .27, p < .005). In addition, the higher stunting (M = 1.2) in the Early Onset versus Late Onset group (M = 0.2, t = 3.99, df = .27, p < .005) is consistent with a greater chronic deficit in linear growth. Although the groups were comparable on growth and developmental parameters at 12 months of age, there were significant differences in attachment classifications in the two groups. In the Early Onset group, 25% (4 of 16) of the children rated as securely attached whereas 83% (10 of 12) of the children in the Late Onset group rated as securely attached (X^2 = 9.3, df = 1, p < .01). At 18 months of age (see Table 12.3), children in the two groups were comparable in physical growth but different in mental development. The Late Onset group had a higher MDI (M = 110.0) than the Early Onset group (M = 95.6, t = 2.79, df = 27, p < 025).

IMPLICATIONS

The present findings indicate that early onset coupled with more chronic FTT is associated with greater impairments in initial physical growth than a later onset.However, the differences between the groups in absolute physical growth are transient and are not manifest at 12 months or at 18 months. On the other hand, early onset and chronicity are associated with differences in psychological variables such as attachment at 12 months and cognitive development at 18 months. The late onset group had a higher frequency of children with secure attachments and higher scores on the Bayley Scale at 18 months than those with early onset. This pattern of findings indicates the potential utility of onset and duration to classify FTT children along dimensions that are relevant to psychological prognosis.

A number of factors may account for differences between the two categories of children. For example, differences may be a consequence of the chronicity of the nutritional deficits associated with poor weight gain. Children with early onset in the present sample experienced undernutrition for greater periods of time than those with later onset and had more impaired physical growth, including head circumference, at study intake. These early differences in physical status may have had longer lasting effects on cognition (Drotar et al., in press). In addition, it is possible that malnourishment may lower the child's social responsiveness (Pollitt, 1969, 1973) and hence contribute to a compromised early attachment. These findings provide empirical evidence for clinical observations of the differences between early versus late onset FTT (Egan et al., 1980; Woolston, 1983). Early onset FTT may represent a distinct subtype that is indicative of more compromised family resources and/or parent-child relationship than later onset FTT. In support of this notion, the early onset group had a greater number of risk factors than the late onset group. For example, the early onset group were from families with significantly lower average incomes ($3,400) than the late onset group ($9,800) ($t$ = 2.63, df = .27, p < .01) and had younger mothers (M = 20.2 vs. 25.2 years) (t = 3.00, df = .27, p < .05). Future research is needed to clearly document the family ecological and biologic risk factors that are associated with early versus late onset FTT and the longer term psychological outcomes of children who differ with respect to onset and/or chronicity.

One potential application of the present findings is that the discrepancy between the onset and identification of FTT may be an important determinant of psychological risk. Given early onset, early identification and treatment may make a difference in psychological risk on salient developmental variables. For this reason, FTT children who are identified and treated early may have better psychological outcomes than those who are treated much later in the course of the condition. In view of the fact that onset and duration may be associated with different psychological outcomes, it will be useful for researchers to specify how their samples differ in onset and duration. Estimates of onset and duration based on physical growth parameters should provide an important complement to objective assessment of physical and psychological status. Researchers might wish to utilize more precise definitions of onset, such as deceleration of rate of weight gain. In addition, it would be useful to determine whether age of onset or chronicity is most critical to subsequent psychological outcome.

The present findings should be replicated to determine generalizability. Sampling characteristics may have influenced the present findings. (Kopp & Krakow, 1982) All children in the sample were hospitalized and received some form of intervention following hospitalization. In addition, the majority of children were from economically disadvantaged families. The present findings may not be as applicable to FTT that occurs in more advantaged families (Chatoor & Egan, 1983). Finally, children were identified very early in life by a research staff that was visible on the pediatric divisions and conducted a chart review of hospitalized children. In practice, many FTT children are not identified as early as those in the present group. On the other hand, it should be possible to identify young infants in pediatric practice who show significant deceleration in rate of weight gain before they have reached the point of severe FTT. Secondary prevention of the longer term psychological deficits associated with FTT may well depend on the efficacy of early identification.

Future research should document psychological outcomes of early onset

FTT children with a more chronic course than the children in this sample. One would hypothesize that children with early onset and long duration (6 months or more) may be at high risk for psychological deficits. The present study underscores the utility of a comprehensive assessment of physical and psychological development outcomes with respect to clinical intervention and follow-up in FTT. Comprehensive outcome studies are very much needed to delineate differential prognosis among children affected with this compelling risk condition.

REFERENCES

Ainsworth, M.D.S., Blehar, M. D., Waters E., & Wall, S. (1978). *Patterns of attachment.* Hillsdale, NJ: Lawrence Erlbaum Associates.

Ainsworth, M.D.S., & Wittig, B. A. (1969). Attachment and exploratory behavior of one year olds in a strange situation. In B. B. Foss (Ed.), *Determinants of infant behavior* (Vol 4). London: Methuen.

American Psychiatric Association. (1980). *Diagnostic and statistic manual of psychiatric disorders* (3rd ed.). Washington, DC: Author.

Bayley, N. (1969). *The Bayley Scales of Infant Development manual.* New York: Psychological Corporation.

Berwick, D. M. (1980). Nonorganic failure to thrive. *Pediatrics in Review, 1,* 265-270.

Bithoney, W. G., & Rathbun, J. M. (1983). Failure to thrive. In W. B. Levine, A. C. Carey, A. D. Crocker, & R. J. Gross (Eds.), *Developmental behavioral pediatrics* (pp. 557-572). Philadelphia: Saunders.

Breunlin, D. C., Desai, V. J., Stone, M. E., & Swilley, J. (1983). Failure to thrive with no organic etiology: A critical review. *International Journal of Eating Disorders, 2(3),* 25-49.

Campos, J. J., Caplovitz, I., Barlett, K., Lamb, M. E., Goldsmith, H. H., & Steinberg, C. (1983). Socioemotional development. In P. H. Mussen (Ed.), *Handbook of child psychology* (pp. 704-915). New York: Wiley.

Chatoor, I., & Egan, J. (1983). Nonorganic failure to thrive and dwarfism due to food refusal: A separation disorder. *Journal of the American Academy of Child Psychiatry, 22,* 294-301.

Drotar, D. (in press). Failure to thrive. In D. K. Routh (Ed.), *Handbook of pediatric psychology.* New York: Guilford.

Drotar, D., & Malone, C. A. (1982). Family-oriented intervention in failure to thrive. In M. Klaus & M. O. Robertson (Eds.), *The Johnson and Johnson Baby Products Company Pediatric Round Table, Birth interaction and attachment* (Vol. 6 pp. 104-112). Skillman, NJ: Johnson & Johnson.

Drotar, D., Malone, C. A., & Negray, J. (1979). Intellectual assessment of young children with environmentally-based failure to thrive. *Child Abuse and Neglect, 3,* 927-935.

Drotar, D., Malone, C. A., & Negray, J. (1980). Environmentally based failure to thrive and children's intellectual development. *Journal of Clinical Child Psychology, 9,* 236-240.

Drotar, D., Nowak, M., Malone, C. A., Eckerle, D., & Negray, J. (in press). Early psychological outcome in failure to thrive: Predictions from an interactional model. *Journal of Clinical Child Psychology.*

Egan, J., Chatoor, I., & Rosen, G. (1980). Nonorganic failure to thrive: Pathogenesis and classification. *Clinical Proceedings of Childrens Hospital National Medical Center, 36,* 173-182.

Elmer, E., Gregg, G. S., & Ellison, P. (1969). Late results of the "failure to thrive" syndrome. *Clinical Pediatrics, 8,* 584–589.

Field, M. (1984). Follow-up developmental status of infants hospitalized for nonorganic failure to thrive. *Journal of Pediatric Psychology, 9,* 241–256.

Fitch, M. J., Cadol, R. V., Goldson, E. J., Jackson, E. K., Swartz, D. F., & Wendel, T. P. (1975). *Prospective study in child abuse: The child study program* (Final Report Office of Child Development Project, Grant No. OCD-CR-371). Denver: Department of Health and Hospitals.

Glaser, H., Heagarty, M. C., Bullard, D. M., & Pivchik, E. C. (1968). Physical and psychological development of children with early failure to thrive. *Journal of Pediatrics, 73,* 690–698.

Gordon, A. H., & Jameson, J. C. (1979). Infant-mother attachment in parents with non-organic failure to thrive syndrome. *Journal of the American Academy of Child Psychiatry, 18,* 96–99.

Hamill, P. V. V., Drizd, T. A., Johnson, C. L., Reed, R. B., Roche, A. F., & Moore, W. M. (1979). Physical growth: National Center for Health Statistics percentiles. *American Journal of Clinical Nutrition, 32,* 607–629.

Hannaway, P. J. (1970). Failure to thrive: A study of 100 infants and children. *Clinical Pediatrics, 9,* 96–99.

Hollingshead, A. B., & Redlich, F. C. (1958). *Social class and mental illness: a community study.* New York: Wiley.

Homer, C., & Ludwig, S. (1980). Categorization of etiology of failure to thrive. *American Journal of Diseases of Children, 135,* 848–851.

Hufton, I. W., & Oates, R. K. (1977). Nonorganic failure to thrive: A long-term follow-up. *Pediatrics, 59,* 73–79.

Kopp, C. B., & Krakow, J. B. (1982). The issue of sample characteristics: Biologically at risk or developmentally delayed infants. *Journal of Pediatric Psychology, 7,* 361–375.

Leonard, M. F., Rhymes, J. P., & Solnit, A. J. (1966). Failure to thrive in infants: A family problem. *American Journal of Diseases of Children, 111,* 600–612.

Lowe, M. (1975). Trends in the development of representational play in infants from one to three years—an observational study. *Journal of Child Psychology and Psychiatry, 16,* 33–47.

Lowe, M., & Costello, A. J. (1976) *Manual for the symbolic play test.* Windsor, Great Britain: N Fer Publishing Co.

MacCarthy, D., & Booth, E. (1979). *Journal of Psychosomatic Research, 114,* 259–265.

Massachusetts Department of Public Health. (1983). *Massachusetts Nutrition Survey.*

Matas, L., Arend, D. A., Sroufe, L. A. (1978). Continuity in adaptation: Quality of attachment and later competence. *Child Development, 49,* 547–556.

Mitchell, W. G., Gorell, R. W., & Greenberg, R. A. (1980). Failure to thrive: A study in a primary care setting, epidemiology and follow-up. *Pediatric, 65,* 971–977.

Money, J., & Annecillo, C. (1976). Change following change of domicile in the syndrome of reversible hyposomatotropinism (psychosocial dwarfism): Pilot investigation. *Psychoneuroendocrinology, 1,* 427–429.

Money, J., Annecillo, C., & Kelley, J. F. (1983). Abuse-dwarfism syndrome: after rescue, statural and intellectual catch up growth correlate. *Journal of Clinical Child Psychology, 12,* 279–283.

Money, J., Annecillo, C., & Werwas, J. (1976). Hormonal and behavioral reversals in hyposomatotropic dwarfism. In E. J. Sachar (Ed.), *Hormones, behavior and psychopathology* (pp. 243–252). New York: Raven Press.

Money, J., Wolff, G., & Annecillo, C. (1972). Pain agnosia and self injury in the syndrome of reversible somatotropin deficiency (psychological dwarfism). *Journal of Autism and Childhood Schizophrenia, 2,* 127–139.

Pastor, D. (1981). The quality of mother-infant attachment and its relationship to toddler's initial sociability with peers. *Developmental Psychology, 17,* 326–335.

Pollitt, E. (1969). Ecology, malnutrition and mental development. *Psychosomatic Medicine, 31,* 193–200.

Pollitt, E. (1973). The role of the infant in marasmus. *American Journal of Clinical Nutrition, 26,* 264–270.

Pollitt, E., & Eichler, A. (1976). Behavioral disturbances among failure to thrive children. *American Journal of Diseases of Children, 130,* 24–29.

Powell, G. F., Brasel, J. A., & Blizzard, R. M. (1967). Emotional deprivation and growth retardation simulating idiopathic hypopituitarism. I. Clinical evaluation of the syndrome. *New England Journal of Medicine, 276,* 1271–1278.

Ramey, C. T., Starr, R. H., Pallas, J., Whitten, C. F., & Reed, V. (1975). Nutrition, response contingent stimulation and the maternal deprivation syndrome: Results of an early intervention program. *Merrill Palmer Quarterly, 21,* 45–55.

Sameroff, A. J., & Chandler, M. J. (1975). Reproductive risk and the continuum of caretaking casualty. In F. D. Horowitz (Ed.), *Review of child development research* (Vol. 4, pp. 187–244). Chicago: University of Chicago Press.

Sander, L. (1964). Adaptive relationships in early mother-child interaction. *Journal of the American Academy of Child Psychiatry, 3,* 231–264.

Schmitt, B. (Ed.). (1978). *The child protection team handbook.* New York: STM Press.

Shaheen, E., Alexander, E., Truskowsky, M., & Barbero, G. J. (1968). Failure to thrive: A retrospective profile. *Clinical Pediatrics, 7,* 225–261.

Sills, R. H. (1978). Failure to thrive: The role of clinical and laboratory evaluation. *American Journal of Diseases of Children, 132,* 967–969.

Silver, H. K., & Finkelstein, M. (1967). Deprivation dwarfism. *Journal of Pediatrics, 70,* 317–324.

Singer, L. T., & Fagan, J. F. (1984). The cognitive development of the failure to thrive infant: A three-year longitudinal study. *Journal of Pediatric Psychology, 9,* 363–383.

Sroufe, L. A. (1979). Socioemotional development. In J. Osofsky (Ed.), *Handbook of infant development* (pp. 462–518). New York: Wiley.

Sroufe, L. A. (1983). Infant-caregiver attachment and patterns of adaptation in preschool: The roots of maladaptation and competence. In M. Perlmutter (Ed.), *Development and policy concerning children with special needs. Minnesota Symposium on Child Psychology* (Vol. 16, pp. 41–84). Hillsdale, NJ: Lawrence Erlbaum Associates.

Waterlow, J. C. (1972). Classification and definition of protein-calorie malnutrition. *British Medical Journal, 3,* 566–569.

Waterlow, J. C., & Rutishauser, I. H. E. (1974). Malnutrition in man. In J. Cravioto, L. Hambraeus, & B. Vahlquist (Eds.), *Early malnutrition and mental development. Symposia of the Swedish Nutrition Foundation* (Vol. 12, pp. 13–26). Stockholm: Almquist & Wiksell.

Waters, E., Wippman, J., & Sroufe, L. A. (1979). Attachment, positive affect and competence in the peer group: Two studies in construct validation. *Child Development, 50,* 821–829.

White, B. L., Kohan, B. T., Antanucci, J., & Shapiro, B. B. (1978). *Experience and environment. Volume 2.* Englewood Cliffs, NJ: Prentice-Hall.

Wolff, G., & Money, J. (1973). Relationship between sleep and growth in patients with reversible somatotropin deficiency (psychosocial dwarfism). *Psychological Medicine, 3,* 18–27.

Woolston, J. C. (1983). Eating disorders in infancy and early childhood. *Journal of the American Academy of Child Psychiatry, 22,* 114–121.

13 Environment and Intelligence: Reversible Impairment of Intellectual Growth in the Syndrome of Abuse Dwarfism

Charles Annecillo
The Johns Hopkins University and Hospital, Baltimore, Maryland

The association of a reversible failure of both statural growth and mental growth associated with child abuse was first ascertained through the study of longitudinal follow-up case records at The Johns Hopkins Hospital (Money, 1977). This syndrome of abuse dwarfism is known also as psychosocial dwarfism, or reversible hyposomatotropinism. Reversible statural and mental growth failure was shown to be typical of the syndrome as a whole, and not only of isolated cases. Rescue from the abusive environment was the decisive determinant for resumption of growth.

The purpose of this chapter is to present a summary of the data on reversible impairment of intellectual growth in the syndrome of abuse dwarfism. In addition to child abuse, other environmental conditions associated with intellectual growth are discussed.

NATURE OF THE SYNDROME OF ABUSE DWARFISM

The syndrome of abuse dwarfism is characterized by a domicile-specific impairment of statural growth and of growth hormone secretion (Patton & Gardner, 1975).Along with various other pathological features of the syndrome, both impairments occur as sequelae of child abuse (Money, 1977). All impairments are reversible upon change of domicile, away from abuse, into an environment of rescue. Taxonomically, the syndrome today is usually known as abuse dwarfism (Money, 1977) or psychosocial dwarfism (Reinhart & Drash, 1969). However, the syndrome has had various diagnostic labels: environmental failure to thrive (Barbero & Shaheen,

1967); deprivation dwarfism (Silver & Finkelstein, 1967); maternal deprivation and emotional deprivation (Powell, Brasel, & Blizzard, 1967; Powell, Brasel, Raiti, & Blizzard, 1967).

Three primary impairments in the syndrome are deficits in physique age, mental age, and social age relative to chronological age. Physique age includes height age, identified as deficient when the height is below the 3rd percentile for chronological age. Physique age includes also the age of onset of puberty, which is delayed if abuse is sufficiently prolonged. Intellectual development is retarded so that the mental age, represented by IQ, is deficient (Money, Annecillo, & Kelley, 1983b). Social maturation also is retarded so that social age, including academic achievement age and psychosexual age, is deficient.

There are three known hormonal impairments in the syndrome. Growth hormone impairment is associated with statural growth impairment resulting in a physique age discrepant with chronological age. Impaired gonadotropin (LH and FSH) secretion is associated with delayed puberty. Impaired response of adrenocorticotropic hormone (ACTH) reserve as tested by metyrapone stimulation is partial—total impairment would be lethal.

The pathognomonic characteristic of the environment of growth failure is child abuse that usually occurs in the parental home where the child is the victim of multiple practices of abuse and neglect. Recovery takes place after rescue. The earlier the rescue, the greater the amount of physical, mental, and social catch-up growth that can be achieved. An initial hospitalization for a short period of time, approximately 2 weeks, allows for the resumption of growth hormone secretion, and leads to the onset of catch-up statural growth, both of which are essential to establish the diagnosis of abuse dwarfism.

Various forms of unusual social behavior represent a developmental deficit reversible upon rescue from abuse. Eating may be from a garbage can, for example, and drinking from a toilet bowl; or there may be binges of polydipsia and polyphagia, possibly followed by vomiting. There can also be a history of such reversible behavioral symptoms as enuresis, encopresis, social apathy or inertia, crying spells, insomnia, eccentric sleeping and waking schedules (Wolff & Money, 1973), pain agnosia and self-injury (Money, Wolff, & Annecillo, 1972), all occurring only in the growth-retarding environment of abuse.

After rescue, impairments such as delayed statural growth and puberty can reverse rapidly whereas other impairments, such as intellectual growth, remain protracted. However, the rate of intellectual catch-up growth is positively correlated with the rate of catch-up statural growth (Money, Annecillo, & Kelley, 1983a). Intellectual catch-up growth has been measured by an increase of as much as 84 IQ points (Money et al., 1983b).

IQ CHANGE AND ENVIRONMENT

IQ Change and the Middle Class

The study on IQ improvement in the syndrome of abuse dwarfism (Money et al., 1983b) is one of a few systematic longitudinal investigations that provide support for the notion that IQ can change and that it changes in association with environmental circumstances.

Constancy versus inconstancy of IQ can be demonstrated only in a longitudinal study by repeated follow-up testing. Major studies of middle-class children have revealed high-positive correlations between IQs tested at two different ages. In general, comparisons of IQs tested at age 18 years with IQs tested at any other age from age 6 years upward show high-positive correlation (Bayley, 1949). These data have been misinterpreted to perpetuate the commonly held belief that IQ remains constant. Although studies reporting IQ change in repeatedly tested normal middle-class children are rare, data from the Fels study (McCall, Appelbaum, and Hogarty, 1973) reveal that IQ does change with wide variation in this population.

McCall et al., (1973) report IQ data on 80 children (38 males and 42 females) who were tested a maximum of 17 times between 2½ and 17 years of age. Despite high-positive correlational stability in IQ between age groups, individual IQ patterns displayed an average longitudinal change of 28.5 IQ points. Shifts of more than 40 points occurred in one in seven children.

Patterns of declining IQs in children were associated with high parental disciplinary penalties. The higher IQ patterns were associated with parental encouragement of intellectual behavior with moderate disciplinary penalties. These parental correlates with IQ change are in agreement with the findings on intellectual growth impairment and recovery in the syndrome of abuse dwarfism wherein abuse is associated with low and declining IQ, whereas rescue in a benign environment is associated with elevation of IQ. However, the typical finding of recovery from severe IQ impairment in the abuse dwarfism study is very different from lack of impairment in the normal middle-class children in the Fels study. The mean IQ for the children in the Fels study was relatively consistent at approximately 118. By contrast, the children rescued from abuse in the abuse dwarfism study had a mean of 66 in abuse and a mean of 90 after various periods of unequivocal rescue (range: abuse, 36–101; rescue: 48–133).

IQ Change, Social Isolation, and Low Socioeconomic Status

Low and presumably declining IQ performance has been reported in various cross-sectional studies of groups of children living in various poor quality environments. Gordon (1923) studied IQ in a group of canal boat children in England who were socially, culturally, and educationally isolated from the general population. He found that the older the child, the lower the IQ; at approximately age 12, the average IQ was 60. Sherman and Key (1932) obtained similar results in a study of isolated Virginia mountain children.

Heber (1976) reported findings on the beneficial effects of an intervention program in black children at high risk for mental retardation born into very low socioeconomic conditions. Focusing upon the birth parents, prevention of mental retardation was achieved through implementation of a vocational and educational intervention program. This form of mental retardation has been labeled cultural-familial (Heber & Garber, 1975) and sociocultural (Heber, 1976), suggesting environmental etiology. Sowell (1977) suggests that the key factor accounting for the lower mean IQ in the black population at large (approximately 15 IQ points below the general population mean of 100) is the disproportionately low socioeconomic status. He provides evidence for low IQ associated with low socioeconomic status among many immigrant groups when they entered the United States. The IQ level of these immigrant groups rose with improvement in their socioeconomic status. It is assumed that IQ scores in the black population would increase with socioeconomic improvement.

IQ Change, Child Abuse, and Neglect

Prior to the 1970s, research dealing directly with an association between child abuse and mental retardation was scarce (Brandwein, 1973). Various studies during the 1970s provide evidence for a deleterious effect of child battering and abuse on intellectual growth (Appelbaum, 1977; Brandwein, 1973; Buchanan & Oliver, 1977; Sandgrund, Gaines, & Green, 1974). The Milwaukee Project (Heber & Garber, 1975) showed further the deleterious effect of the neglect of developmental stimulation and the beneficial effect of intervention in preventing developmental mental retardation.

Constancy versus inconstancy of IQ can be demonstrated only in a longitudinal study, that is, by repeated follow-up testing. Comprehensive longitudinal investigation of intellectual impairment in institutionalized infants and children can be found in studies conducted by Skeels (1966) and Dennis (1973). The environments of intellectual growth retardation within the institutions in both studies warranted the description of institutional abuse and neglect. In brief, they were overcrowded and inadequately staffed. Each author's descriptions of their respective institutional environments are very similar and reminiscent of the more well known institutional conditions described by Spitz (1946).

In 1966, Skeels published his outcome study of mental retardation as a sequel to infantile institutional abuse and neglect (Skeels & Fillmore, 1937). A study group of children was transferred to an enriched institutional environment where they subsequently displayed dramatic intellectual growth improvement as compared to a group of comparable children who continued in their regular living arrangement. Adult follow-up revealed various qualitative differences between the groups. The study group was far superior in educational and vocational achievement and was found to be leading a more normal independent life style. Dennis (1973) further demonstrated intellectual impairment as a sequel to institutional abuse and neglect. He found that the younger the institutionalized foundlings were when adopted into normal family life, the earlier the resumption of normal intellectual growth and the higher the ultimate level of the adult IQ.

IQ DATA IN THE SYNDROME OF ABUSE DWARFISM

Sample and Procedures

The sample comprised a group of 34 patients, 15 females and 19 males (27 white and 7 black), with a diagnosis of abuse dwarfism.Except for 4 patients at around the 4th percentile, all were below the 3rd percentile in stature. All of the patients had a history of abuse documented in their medical and social service records.

The patients were all seen and evaluated in the Psychohormonal Research Unit and the Pediatric Endocrine Clinic at The Johns Hopkins Hospital during the period 1958–1975. From a total of 50 patients, 34 met the criteria of having at least one follow-up IQ testing for before and after rescue comparison. Some patients had more than one follow-up IQ test and more than one change of environment.

Each patient has a consolidated case history on file in the Psychohormonal Research Unit. Most of the intelligence testing for this study was conducted by members of the Psychohormonal Research Unit. Relevant information was abstracted and reduced for each patient in chronological order of IQ testing and history of abuse and rescue. Ratings of the patient's domiciliary environmental conditions determined whether abuse was present, equivocal, or absent.

IQ Change in Consistent Rescue

Table 13.1 shows the results of IQ change in 23 patients who had a baseline abuse IQ and a final follow-up IQ after varying periods of consistent rescue. As a group, these patients displayed a significant IQ increase (t test for correlated samples, $z = 5.5$, $p < .001$).

When children with the syndrome of abuse dwarfism are rescued, their rate of statural growth accelerates, and they go through a period of catch-up growth. The change in rate of mental growth, if similar to that of statural growth, would be reflected in a progressive increase in IQ. A sudden increase in IQ would, by contrast, signify a response to a sudden lifting of some sort of inhibiting constraint on the manifestations of intelligence. If the hypothesis of progressive catch-up intellectual growth is correct, then there should be a correlation between the amount of IQ increase and the amount of time spent free of abuse after having been rescued.

Duration of rescue and two other variables relevant to resultant IQ, baseline IQ, and variability in age, were subjected to a multiple regression analysis. The analysis revealed that these differences together account for 60% of the variance in IQ elevation ($r = .779$, $p < .005$). Time in rescue accounts for most of the variance (z beta $= 4.47$) as compared with baseline IQ level (z beta $= 3.02$), while baseline age acted as a moderator variable with a negative beta weight (z beta $= -2.59$). The moderating role of age means that the younger the age at the time of rescue, the greater the amount of IQ elevation during a comparable period of time after rescue. Overall, the regression analysis supports the hypothesis that postrescue growth of intelligence is not sudden, but progressive over time.

IQ Change in Younger Versus Older Age Groups

In order to test further the age hypothesis, namely, that the younger the age at the time of rescue, the greater the amount of intellectual catch-up growth and IQ elevation, two subsets of patients were assembled (Table

TABLE 13.1
IQ Elevation After Rescue ($N = 23$)

IQ	Before Rescue	After Rescue	Increase in IQ	Age Before Rescue	Age After Rescue	Increase in Age
Mean	66	90	24	7-7	12-8	5-1
SD	16	21	21	4-4	5-11	3-1

Note. Age in years and months.
$R = .78$; $p < .005$

13.2). The table shows a mean elevation of IQ in each subset (33 and 16, respectively) which, when combined, attained a high degree of significance ($F = 27.18, p < .001$). Though this significance applies regardless of age of rescue, it is also evident that the younger the age at rescue, the greater the gain in IQ. Analysis of variance showed that the difference between the two subsets reached significance at the level of $p < .1$ ($F = 3.46$), which in view of the small size of the sample is quite substantial. In addition, the effect of age at the time of testing reached significance at the level of $p < .05$ ($F = 5.39$), reflecting rather convincingly the fact that the younger patients had, in general, higher IQs. Because they were younger, one may infer, they had suffered less IQ impairment prior to being rescued and therefore were able to benefit more from catch-up growth.

IQ Change in Equivocal Versus Abuse or Rescue Environments

Some patients spent at least one period of their lives in conditions that qualified as equivocal, that is, as definitely neither abusive nor nonabusive, or an alternation of both.If they were IQ tested at the onset and conclusion of this period, regardless of how many other tests they had, then they qualified for inclusion in Table 13.3.

Table 13.3 shows a consistent trend confirming IQ impairment in an environment of abuse and IQ elevation after change to an environment of rescue. The new finding of this table is that after removal from the environment of abuse, even when the change was rated as equivocal and not as a full rescue, IQ deterioration ceased or showed a minor degree of elevation.

Statistical evaluation of Table 13.3 required two analyses of variance — one comparing the three environmental sequences that began with abuse, and the other comparing the three sequences in which the second condition was unchanged from the first. In both analyses, the main effect for trials was significant at the level of $p < .01$ ($F = 12.52$ and 6.12, respectively), despite the deviant effect of the abuse/abuse group. Correspondingly, in

TABLE 13.2
Post-abuse Elevation of IQ Relative to Age at Time of Rescue

N	Age at Rescue Range	Mean	Rescue Mean	Baseline IQ $\bar{x}$ & SD	Follow-up IQ $\bar{x}$ & SD	IQ Elevation $\bar{x}$ & SD
7	2-4 to 5-5	3-9	3-11	71 ± 21	104 ± 11	33 ± 24
7	5-8 to 15-7	10-3	4- 1	63 ± 15	78 ± 16	16 ± 7

Note. Age in years and months.

TABLE 13.3
Change in IQ Relative to Abuse, Equivocal and Rescue Environments

Row	Environmental	N	$\bar{x}$ Years of Test-test Interval	$\bar{x}$ IQ Baseline	$\bar{x}$ IQ Followup	$\bar{x}$ IQ Diff.
1	Abuse/Abuse	5	8-0	83 ± 16	75 ± 4	−8 ± 15
2	Abuse/Equivocal	6	4-11	61 ± 9	78 ± 7	17 ± 8
3	Abuse/Rescue	16	4-11	67 ± 17	92 ± 16	24 ± 18
4	Equivocal/Equivocal	8	4-6	75 ± 12	82 ± 13	7 ± 10
5	Rescue/Rescue	12	4-2	77 ± 17	96 ± 19	19 ± 13

Note. $N = 34$, but 13 patients qualified for inclusion in two different sequences because of multiple changes of domicile over widely spaced periods of their lives.
Analysis of variance:
Rows 1, 2, 3: trials ($F = 12.52, p < .01$)
            interaction ($F = 3.43, p < .05$)
Rows 1, 4, 5: trials ($F = 6.12, p < .01$)
            interaction ($F = 3.55, p < .05$)

both analyses the interaction effect between IQ change from one environment to the other was significant at the level of $p < .05$ ($F = 3.43$ and 3.55, respectively).

A post hoc test of simple main effects using a pooled error term showed that the interactions were largely the result of the dramatic improvements shown by the abuse/rescue and rescue/rescue groups. These data yet again confirm the impairment or deterioration of intellectual growth in association with abuse environments and improvement of intellectual growth in rescue, even when rescue is equivocal.

CONCLUSION

There are factors implicit in the social environment that profoundly influence intellectual growth without altering the level governed by genetic make-up. Although the genetic make-up probably sets the upper limit for intellectual growth, the achievement of optimum intelligence is possible only under ideal environmental circumstances. Modern genetic theory avoids dichotomizing the genetic versus the environmental and "postulates a genetic norm of reaction which, for its proper expression, requires phyletically prescribed environmental boundaries" (Money & Ehrhardt, 1972). Child abuse constitutes one environmental factor that constricts the prescribed environmental boundary and inhibits intellectual growth. Rescue from abuse releases the inhibition and widens the boundary. The understanding that genetic make-up does not exercise exclusive determining power over the establishment of IQ, and the knowledge that environmental circumstances can facilitate or inhibit intellectual growth, are essential to professionals involved in the care and rescue of children who are victims of institutional and noninstitutional child abuse and neglect. The findings in the abuse dwarfism study demonstrate that by manipulating the environmental circumstances of affected children, permanent mental retardation can be prevented. In addition, substantial intellectual catch-up growth can occur in rescue even when rescue from abuse is equivocal or not achieved until the adolescent years.

REFERENCES

Appelbaum, A.S. (1977). Developmental retardation in infants as a concomitant of physical child abuse. *Journal of Abnormal Child Psychology, 5,* 417–423.

Barbero, G. J., & Shaheen, E. (1967). Environmental failure to thrive: A clinical view. *Journal of Pediatrics, 71,* 639–644.

Bayley, N. (1949). Consistency and variability in the growth of intelligence from birth to eighteen years. *Journal of Genetic Psychology, 75,* 165–196.

Brandwein, H. (1973). The battered child: A definite and significant factor in mental retardation. *Mental Retardation, 11,* 50–51.

Buchanan, A., & Oliver, J. E. (1977). Abuse and neglect as a cause of mental retardation: A study of 140 children admitted to subnormality hospitals in Wiltshire. *British Journal of Psychiatry, 131,* 458–467.

Dennis, W. (1973). *Children of the Creche.* New York: Appleton-Century-Crofts.

Gordon, H. (1923). Mental and scholastic tests among retarded children. *Education Pamphlet 44.* London: Board of Education.

Heber, F. R. (1976, June). *Sociocultural mental retardation—A longitudinal study.* Paper presented at the Vermont Conference on the Primary Prevention of Psychopathology.

Heber, F. R., & Garber, H. (1975). The Milwaukee Project: A study of the use of family intervention to prevent cultural-familial mental retardation. In B. Z. Friedlander, G. M. Sterrit, & E. K. Girvin (Eds.) *Exceptional infant: Vol. 3. Assessment and intervention* (pp. 399–433). New York: Brunner/Mazel.

McCall, R. B., Appelbaum, M. I., & Hogarty, P. S. (1973). Developmental changes in mental performance. *Monographs of the Society for Research in Child Development. 38* (3, Serial No. 150).

Money, J. (1977). The syndrome of abuse dwarfism (psychosocial dwarfism or reversible hyposomatotropinism): Behavioral data and case report. *American Journal of Diseases of Children, 131,* 508–513.

Money, J., Annecillo, C., & Kelley, J. F. (1983a) Abuse-dwarfism syndrome: After rescue, statural and intellectual catchup growth correlate. *Journal of Clinical Child Psychology, 12,* 279–283.

Money, J., Annecillo, C., & Kelley, J. F. (1983b). Growth of intelligence: Failure and catchup associated respectively with abuse and rescue in the syndrome of abuse dwarfism. *Psychoneuroendocrinology, 8,* 309–319.

Money, J., & Ehrhardt, A. A. (1972). *Man and woman, boy and girl: The differentiation and dimorphism of gender identity from conception to maturity.* Baltimore: Johns Hopkins Press.

Money, J., Wolff, G., & Annecillo, C. (1972). Pain agnosia and self-injury in the syndrome of reversible somatotropin deficiency (psychosocial dwarfism). *Journal of Autism and Childhood Schizophrenia, 2,* 127–139.

Patton, R. G., & Gardner, L. I. (1975). Deprivation dwarfism (psychosocial deprivation): Disordered family environment as cause of so-called idiopathic hypopituitarism. In L. I. Gardner (Ed.), *Endocrine and genetic diseases of childhood and adolescence* (2nd ed., pp. 85–98). Philadelphia: Saunders.

Powell, G. F., Brasel, J. A., & Blizzard, R. M. (1967). Emotional deprivation and growth retardation simulating idiopathic hypopituitarism. I. Clinical evaluation of the syndrome. *New England Journal of Medicine, 276,* 1271–1278.

Powell, G. F., Brasel, J. A., Raiti, S., & Blizzard, R. M. (1967). Emotional deprivation and growth retardation simulating idiopathic hypopituitarism. II. Endocrinologic evaluation of the syndrome. *New England Journal of Medicine, 276,* 1279–1283.

Reinhart, J. B., & Drash, A. L. (1969). Psychosocial dwarfism: Environmentally induced recovery. *Psychosomatic Medicine, 31,* 165–171.

Sandgrund, A., Gaines, R. W., & Green, A. H. (1974). Child abuse and mental retardation: A problem of cause and effect. *American Journal of Mental Deficiency, 79,* 327–330.

Sherman, M., & Key, C. B. (1932). The intelligence scores of isolated mountain children. *Child Development, 3,* 279–290.

Silver, H. K., & Finkelstein, M. (1967). Deprivation dwarfism. *Journal of Pediatrics, 70,* 317–324.

Skeels, H. M. (1966). Adult status of children with contrasting early life experiences: A followup study. *Monographs of the Society for Research in Child Development 31* (3, Serial No. 105).

Skeels, H. M., & Fillmore, E. A. (1937). The mental development of children from underprivileged homes. *Journal of Genetic Psychology, 50,* 427–439.

Sowell, T. (1977, March 27) New light on black IQ. *New York Times Magazine,* pp. 56–62.

Spitz, R. A. (1946). Hospitalism. *Psychoanalytic Study of the Child, 2,* 113–117.

Wolff, G., & Money, J. (1973). Relationship between sleep and growth in patients with reversible somatotropin deficiency (psychosocial dwarfism). *Psychological Medicine, 3,* 18–27.

14

Size Versus Age: Ambiguities in Parenting Short-Statured Children

Diane L. Rotnem
Yale University School of Medicine, New Haven, Connecticut

Ambiguities in the parenting of children whose size is discrepant with chronological age remains one of the most perplexing challenges in the upbringing of short-statured children. Inherent in such ambiguities is a tendency for the environment (i.e., parents, siblings, peers, etc.) to relate to the short-statured child according to size rather than chronological age. Not uncommonly, this leads to infantilization and lower developmental expectations than those for normal size peers (Rotnem, Cohen, Hintz, & Genel, 1979; Rotnem, Genel, Hintz, & Cohen, 1977). The extent to which this tendency interferes with achievement of age and phase-appropriate developmental tasks constitutes a major psychosocial risk of short stature (Stabler, Whitt, Moreault, D'Ercole, & Underwood, 1980). This chapter describes some of the fundamental dilemmas faced in the parenting of short-statured children. It is meant to provide health care professionals with a context in which to think about helping short-statured children and their families to develop more effective coping strategies.

There is general agreement that the most critical variable in the psychosocial adjustment of short-statured persons is parental functioning and decision making in regard to promotion of age and phase-appropriate developmental tasks (Drash, 1969; Rotnem et al., 1979). However, the effect of parental attitudes and coping styles on the social and psychological competence of short-statured children has not been fully explored, nor have the important questions in this regard been systematically studied. Most of our data are anecdotal at best, gleaned from interviews and observations of parents and children.

CONTRIBUTIONS OF THE
PREEXISTING PARENT–INFANT RELATIONSHIP

The subject of parenting is broad and ambiguous in and of itself.The complexities of parent–child interaction are rooted in earliest infancy and in the attachment process which begins prenatally (Bibring, Dwyer, Huntington, & Valenstein, 1961). The infant and its caregiving environment have been described as a biological system, fully interactive, even before birth (Sander, 1981). The meaning of the particular child to each parent, the motivation for the pregnancy, expectations the parents have of the child, general health of the infant, and so forth, all influence the attachment or "claiming" process.

A critical component of the child's contribution to the reciprocity of the parent–infant relationship is his or her own temperament and the extent to which she or he is responsive to parental efforts to comfort. The capacity for empathic attunement with the infant on the part of the parent, and individual personality characteristics of each parent, are major contributions to the evolving relationship (Emde & Sorce, 1983). Through their temperament and developmental characteristics, together called endowment, children actively shape and elicit certain handling styles and feelings from their parents (Greenspan, 1981). Pruett describes the parental caregiving system as comprising a complex mosaic of interactional systems between mother and child, between father and child, between the marital dyad and the child, and between the child and the relationship between mother and father (Pruett, 1985). He suggests that fathers, as well as mothers, are capable of powerful parental feelings and "mothering behavior." It is in this context that complex environmental influences shape the short child's developing sense of self.

INTERACTION BETWEEN SELF
AND THE ENVIRONMENT

Thus it is that the child's self-concept has roots in earliest childhood experiences, in interactions with parents, grandparents, siblings, extended family, friends and community.Together, these provide the child with a sense of identity including an ability to discriminate between self-expectations and those implied or expressed by others. For children, the concept of personal identity develops gradually and in close connection with the development of physical characteristics, e.g., size, strength, gender, functioning of the body (Erikson, 1959). Parental perception of those characteristics significantly affects the interaction (Cohen, Dibble, & Grawe, 1977; Winnicott, 1965).

Even normal physically well proportioned children, although they cannot ignore the reality of being smaller than their parents or older siblings, require guidance in reaching goals. However, they usually do not feel small or inadequate because their goals are realistically geared to their competence level. They succeed far more than they fail, and they are generally accepted within a peer group. Short children often elicit response from the environment appropriate for a much younger child. Their small size attracts attention. The major developmental task facing the short child becomes one of adaptation to such an environment. This process begins with parents whose attitudes toward their short child significantly influence the child's concept of self. Thus, the developmental patterns of short-statured children may be best understood within the context of their parenting and social experience. For example, children who regard their bodies as unacceptable in some way may be so preoccupied with that concern that they cannot concentrate on other things, such as learning, social interactions with peers, or physical activity. This may lead to withdrawal from a peer group, sensitivity to rejection, cognitive inhibition, and eventually to inadequate socialization.

SHORT STATURE AS A CHRONIC CONDITION

Regardless of the etiology of the short stature, children and families must be prepared realistically to face years of medical supervision and concern about psychosocial adjustment.Furthermore, despite resistance to thinking of short stature as a disease entity, parents of children with growth delay are united by experiences they share in common with all parents who face the threat of serious illness or disability in their child. Uncertainties about the future loom over these parents, including difficult decisions regarding treatment.

Like parents of children with chronic diseases, parents of short children may find themselves living with the uncertainty between hope and despair. Their lifestyle may be significantly altered by the time and expense required by prolonged medical supervision. Every relationship within the family may be influenced by the stresses and uncertainties associated with a child whose growth is delayed.

Children whose health is impaired are nonetheless maturing and changing. Whether their disorder is long-standing or newly acquired, it will have special emotional, intellectual, and social meaning for the children and their families at each successive stage of their development. New issues will continue to arise, and these will challenge the adaptive capacities and coping skills of the children, their parents, and their siblings. The burden is on parents to assist the children in developing strategies for coping with short stature, as well as strategies for approaching future expectations realistically. The responsibility for assisting parents in these difficult tasks lies with health care professionals and educators.

There are several parallels between parents of short children and those with children who have other chronic diseases. First, unless the family has a history of short stature, the diagnosis may come as a shock to both parents and child. The parents may be upset at the first realization that their child is not growing. Often, the earliest diagnostic indicator is an intuitive sense that parents may have about their child. Perhaps a tangible marker such as a failure to outgrow clothing from one season to the next is the first indicator. Parents commonly struggle with a sense of isolation in facing the reality that their child is not growing, or is different from other children. They struggle with the question of who should be told—spouse, the child, grandparents, pediatrician, and so on. They struggle with a wish to deny the problem versus a wish for resolution. They also struggle with feelings of responsibility for causing the short stature.

Second, they struggle with the timing of when to share their concerns with others. There is often a struggle with friends and relatives, even well-meaning professionals who reassure, "It's too soon to be worried," "Let's wait another 6 months and see where we are then." How can parents make a decision to pursue an evaluation in the face of such opposition expressed in the form of reassurance? What are the risks of waiting? What are the social-psychological risks of treating versus not treating?

Parents enter into a process around the time of the initial diagnosis that includes the following features characteristic of parents with chronic disease children:

The Diagnostic Phase.   A careful diagnostic evaluation to establish the etiology of the condition may help parents as well as the children with a realistic understanding of the causes, and what they might expect from treatment. What is absolutely predictable is the heightened anxiety for parents and children around the time of the diagnostic evaluation. Some parents find it difficult to accept the diagnosis and seek additional opinions, hoping to hear a dissenting opinion or a reversal of the diagnosis. Similarly, wishes for a "magical cure" may lead to heightened anticipation and increased anxiety. Health caregivers who tend to see parental anxiety as interfering with their medical management may feel angered and unable to respond to the appropriate level of concern expressed in the anxiety. Other parents may need to deny that their child is different, and may deny their child is like other short-statured persons. For most parents, the discrepancy between size of their child and the child's peers will become increasingly apparent with age. Parents whose denial is interfering with the recommended medical regimen are among those who benefit most from informal or group contacts with other parents and families who have confronted similar feelings.

Coming to terms with the reality of the painful truth of the diagnosis of growth delay mobilizes defenses such as avoidance, denial, intellectualization, and projection. Thus, the evaluation and facing the diagnosis begin a process of grieving the loss of the "wished-for perfect child" (Solnit & Stark, 1961). Health care professionals may facilitate that process by helping the parents to express their feelings about the diagnosis with special attention to the parents' needs to express their disappointment, without guilt, over the "less than perfect child."

A period of disbelief accompanied by anger, denial, avoidance, or intellectualization is common. A parental tendency to deny the short stature often occurs in cases where other major problems exist, and short stature is considered to be perhaps the least of the child's difficulties. With most forms of short stature, children are physically well proportioned, with no unusual physical stigmata except for their short height. As a result, it may be easier for parents to deny the need for medical attention.

Parents may believe that the professionals are mistaken about their child. They may be unable to process what is being said to them, especially about a new diagnosis. They often cannot remember what was said, much less convey the information to family members or friends. Resistance to believing the diagnosis, or thinking and talking about it without emotion, are common responses that can either help or hinder parents in coming to terms with their child's condition. It is crucial that health care professionals understand the meaning of individual parental coping styles, particularly their use of defenses. "Forgetting" is a common defense against emotional pain, as well as the flood of medical information, particularly if the latter is delivered in a manner that does not allow for dialogue and "working through." Parents often view forgetting what the doctor said as a sign of their own inadequacy, and feel shy and apologetic about asking the doctor to repeat what has already been explained. The result is one of our greatest fears for parents and children—that they do not understand. Children need to be fully informed about the condition for which they are being treated. Realistic expectations about treatment have influenced psychosocial adjustment.

Grief and Disappointment.   When children develop an illness or condition that in some way threatens their bodily integrity, grief is inevitable for those who care about them (McCollum, 1975). Although growth delay is usually not life threatening, a parent grieves the loss of the "wished-for perfect child," including the loss of certain hopes and dreams (Solnit & Stark, 1961). Some hopes will need to be modified; some must be relinquished altogether. Grief can be isolating. Parents without acceptable channels for the expression of their grief and disappointment may find these feelings interfering with their relationship with the child, e.g., reinforcing feelings of unacceptability or inadequacy in the child.

Feelings of Inadequacy and Guilt.   All parents of children with health impairment experience guilt. Feelings of responsibility for the child's growth failure, though irrational, are real. Chronically guilt-laden parents are those who are unable to like or respect themselves, and this has direct implications for the kinds of identification children form with both parents. These parents will have difficulty helping their child to maintain his or her own feelings of worth and dignity (McCollum, 1975). These parents may seek ways to atone for their guilt by lavishing excessive attention and care upon the child (who is almost certain to become aware that he or she is a burden). Such over compensation is a well-recognized parental coping style with children who are subjected to multiple medical procedures and hospitalizations. The consequences of this particular coping style on the part of the parents, grandparents, and other family members are well known to health care professionals and educators who may have a tendency to view these children as "spoiled" or "manipulative."

Guilt Arising from Negative Feelings.   Even the kindest and most conscientious parents have cause at time to feel irritation, disappointment, resentment, or anger toward their children. Parents of healthy children can usually accept these feelings and still feel adequate as parents (McCollum, 1975). However, parents of children with chronic conditions such as growth delay tend to be more troubled by negative feelings toward their children (McCollum, 1975). In fact, when a sense of remorse or feelings of worthlessness persist it may be that there are some unacknowledged angry feelings toward the child that require exploration (Ferholt et al., 1985).

In conditions of growth failure that are believed to be more environmentally induced (e.g., failure to thrive, psychosocial dwarfism), intense parental ambivalence and antagonism play the prominent role in the growth failure and secondary depression. Perhaps most noxious for these children is the affective withdrawal of parental investment, which the child experiences as a loss of a crucial affective tie (Ferholt et al., 1985).

Anger.   Anger generated from a sense of helplessness about coping with the problem of growth delay may express itself in many forms including irritability within the family, displaced feelings of anger deflected from the child onto the spouse and vice versa, critical feelings toward professionals, bitterness, marital conflict, apprehension, and excessive anxiety. Identification of such behaviors warrants examination of the fears and underlying sense of helplessness felt by parents of children with growth delay. Parents with adequate support systems are better able to explore and understand their negative feelings.

Guilt and Overprotection. Parker and Lipscombe (1981) have attempted to identify the key constructs of parental overprotectiveness, a parental coping style often associated with the parenting of short children. Maternal overprotection has been associated with several neurotic and psychotic disorders as identified retrospectively from adult psychiatric populations (Parker, 1983). However, little is known about determinants of such parental characteristics or the extent to which it is operating in the parenting of short-statured persons. It is commonly observed that in their wish to spare their child the embarrassment of teasing, parents of short statured children they may intervene more frequently on behalf of the child. This parental behavior might be interpreted as a tendency toward overprotection, which may have the undesirable effect of thwarting the child's efforts to be self-reliant. In our own study, 80% of parents acknowledged feelings of guilt, and a resultant tendency toward behaviors that might be interpreted as overprotective, intrusive or excessively controlling. These parental coping styles had the effect of inhibiting the development of the child's capacity for self-reliance and mastery in coping with anxiety-producing situations.

Perception of the Child as Vulnerable. Parents of children in our studies generally agreed that they perceived their short child as more vulnerable than normal-statured children. Difficulties in limit setting, consistency, and age-appropriate expectations were dilemmas expressed as a result of the perceived vulnerability in the child (Rotnem et al., 1977; Rotnem et al., 1979). According to most parents' view, the short-statured child's difficulties fall into four areas: (1) feeling smaller and less competent than younger siblings; (2) teasing by peers (e.g., name calling, scapegoating, avoidance); (3) wishing to be babied; (4) reactions of others in public places. Most parents express a wish for developmental guidance regarding the rearing of their children.

Short-statured children constitute a group vulnerable to developmental disturbances. They are particularly vulnerable to difficulties in peer groups. In our experience, however, the quality of parenting can make the critical difference in the extent to which competence is emphasized by the parents in promoting mastery of age- and phase-appropriate developmental tasks. For example, although most vulnerable to developmental difficulties in mid- and late childhood, these children seem to have potential strength for coping, provided they have had relatively normal early caregiving experience (Rotnem et al., 1977; Rotnem et al., 1979). The effects of early caregiving experience on the child and family's adaptation to the growth disorder have not been adequately studied. However, some assessment of qualities that characterized the parent–child relationship prior to the identification of the problem may provide important indicators for prediction of ultimate psychosocial adjustment to the short stature.

Parental Perception of Child's Size.   Parents' perception of their child's size is significant in determining expectations (Grew, Stabler, Williams, & Underwood, 1983). Because the response evoked by the visual stimulus of a short child is generally a tendency to relate to the child as much younger, parents, health care professionals, and educators must constantly remind themselves of the child's chronological age. In our study, children who were mastering age-appropriate emotional tasks had parents who fostered their sense of competence (Abbott, Rotnem, Genel, & Cohen, 1982). Effectively functioning children did not elicit the same response from hospital personnel as the less competent children, who had a tendency to be more infantilized and to accept the infantilization (Rotnem et al., 1979). Denial, provided it does not interfere with the medical regime, may be a positive adaptation to the problem if it encourages age-appropriate behavior.

Impact of Short Stature on Family Functioning.   The impact of short stature on parental and family functioning has not been studied adequately. Although short children tended to view themselves as the smallest of the children in their families, they generally indicated feelings of belonging and acceptance within their families. This was in contrast to their most striking sense of isolation within their peer group (Abbott et al., 1982). For example, on projective testing, almost all could perceive a fight with a peer, but tended to respond with sadness or fear, rather than assertiveness or retaliation.

The disruption on family functioning brought about by a child who requires extensive medical and perhaps psychological evaluation and close following is observed in families with any kind of health impairment.

Sibling Reactions.   One of the greatest challenges presented to parents of children with growth delay is facilitating positive sibling interactions. As with any chronically ill child who requires time from the parent, siblings have feelings of competition with the parent for equal time. This is compounded by the fact that the short-statured child may indeed be less competent than his or her younger siblings. Potential for scapegoating within the family must be monitored carefully, as siblings search for ways of expressing their aggressive feelings toward the short-statured child. A major factor is the extent to which parental attitudes toward the short-statured child differ from those toward his or her average size siblings.

IMPACT OF SHORT STATURE
ON THE CHILD AND FAMILY

The threats to bodily integrity experienced by the chronically ill child are similar regardless of which organ system the disorder affects (Burlingham, 1972).The short-statured children in our study tended to be self-critical and to blame themselves for their short stature. Their major concern was "What if I don't grow?" as if holding themselves personally responsible. The use of denial as a defense was inadvertently encouraged by parents, pediatricians, and teachers, who tended to express the hope that the short child would have a spurt of growth when he or she got older. Parents frequently feel failure about their competence because their child fails to grow. This conveys to the short child the sense of being unacceptable or incomplete. It depreciates the meaning of the present and directs the child's attention to the future when a magical process of "growing older" would also mean "growing bigger." Thus, recognition of the child's handicap by the family and sharing of concern were sometimes inhibited by a "let's keep hoping" attitude. Many primary care physicians also shared in this wishful thinking; consequently, children with growth hormone deficiency often had a prolonged delay in diagnosis.

Despite the best intentions, size and appearance powerfully shape parental expectations. Parents in our studies tended to underestimate their children's emotional and developmental difficulties. Their standards of behavior were tailored to the child's size and appearance, not necessarily emotional needs. The guilt of many parents with a physically handicapped child often found expression through over-protectiveness and excessive control (Love, 1970). This parental style further limited the child's capacity for developing self-reliance and autonomy.

Pre- and early adolescence may be a particularly difficult period for the short-statured child. The support of parents is even more critical for the short-statured child who cannot find encouragement from a peer group as readily as normal children. At the same time, autonomous stirrings in the adolescent make parental support less available and useful. Short-statured children do not always have significant developmental problems, particularly if the family makes a conscious effort to foster the child's sense of competence. In our studies, effectively functioning short-statured children, who were of the same general size and appearance as the other hypopituitary children, did not seem to elicit the same kind of age-inappropriate behavior from hospital personnel. Through their greater personal maturity and no-nonsense directness, they made it apparent that they neither expected, nor would accept, infantilization. Their behavior overrode the effect of their appearance in shaping social interactions (Abbott et al., 1982; Rotnem et al., 1977).

INTERVENTION

Short children should be considered vulnerable to developmental distur-bances (Anthony & Koupernik, 1974; Brust, Ford, & Rimoin, 1976; Rotnem et al., 1977; Rotnem et al., 1979). To reduce this vulnerability, early parental guidance and other forms of supportive psychotherapeutic proce-dures may be useful, as well as a coordinated team approach to the care of these children and their families (Meyer-Bahlburg, 1985). However, the ultimate benefit of comprehensive approaches to the care of short children remains to be determined by longitudinal studies.

Because the school is the primary socializing environment for children once they reach school age, it should be a target environment for interven-tion (Holmes, 1982). Educational planning is still sometimes based on judgments about the child's size. Size, however, should not be a reason for retention. Teasing can lead to exclusion or withdrawal from peer group activities and eventually to social withdrawal. Public facilities such as schools are still designed for normal or even tall-statured individuals; thus it is, that in school the child is likely to encounter real logistical problems, such as the height of the school desk or a blackboard. Restrooms, door handles, and elevator buttons are additional obstacles encountered on a day-to-day basis by the short-statured person to which normal-statured people are oblivious.

One model of integrated medical and psychosocial care comes out of the Syracuse group (Finley, Crouthamel, & Richmond 1981). Its compo-nents are the following:

1. Complete medical and psychological evaluation and diagnosis of the child and family.
2. Coordination of the patient's total treatment program.
3. Assistance in developing healthy coping mechanisms, a positive self-image and satisfying peer relationships and realistic educational and vocational goals.
4. Patient and parent education regarding the growth disorder and its treatment in an attempt to reduce anxiety in the child and family.
5. Educational outreach through mental health care students; and coordination of medical center efforts with community agencies and referring physicians.

Although it is generally accepted that the psychosocial outcome for adequately supported parents and children is significantly more hopeful, there have been no systematic studies designed to look at outcome for those children and families. In addition to help from professionals, parents can derive support from parent support groups at the local, state, and national levels (Ablon, 1978). Organizations such as the Human Growth Foundation and Little People of America provide a network of mutual help for short-statured persons and their families.

Little People of America groups have emphasized a cognitive approach to restructuring of the self-image. This involves social destigmatization through confrontation. This group advocates that the physical identification of being a dwarf is effective in altering various aspects of social interaction for dwarfed persons (Ablon, 1984; Weinberg, 1968). To measure the relative efficacy of such programs, intervention studies that will yield both short-and long-term outcome data are needed. Measurement of the effect of treatment in terms of cost effectiveness, adult social integration, family functioning, compliance with medical regime, and school adjustment. Contact with other short children may be useful in reassuring the child that he or she is not alone.

SUMMARY

Who is the patient? The patient is not only the identified child, but is the entire family. The challenge to parents is to support the child in developing effective strategies for coping with social influences resulting from growth delay. The child's adaptation to his or her size and the limitations that may impose on his or her life is largely determined by the capacity of parents to come to terms with their own grief and disappointment over the less-than-perfect child, and to promote the child's mastery of age- and phase-appropriate developmental tasks. Early detection of growth delay will provide for earlier attempts at therapeutic intervention (Aynsley-Green & MacFarlane, 1983). The support and communication of acceptance may be more critical for short-statured than normal-sized children because the former are less likely to find encouragement from a peer group. Outcomes of treatment should be measured more in terms of the degree of social integration achieved in adult life than in terms of height achieved. Approaching the treatment as a long-range collaboration between children, parents, and professionals will do much to ensure that psychosocial growth keeps pace with physical growth.

REFERENCES

Abbott, D., Rotnem, D., Genel, M., & Cohen, D. J. (1982). Cognitive and emotional functioning in hypopituitary short-statured children. *Schizophrenia Bulletin, 8,* 310–319.

Ablon, J. (1981). Dwarfism and social identity: Self-help group participation. *Social Science and Medicine. 15B,* pp. 25–30.

Ablon, J. (1984). *Little people in America: The social dimensions of dwarfism.* New York: Praeger.

Anthony, E. J., & Koupernik, C. (Eds.). (1974). *The child in his family, 3.* New York: Wiley.

Aynsley-Green, A., & MacFarlane, J. A. (1983). Method for earlier recognition of abnormal stature. *Archives of Diseases of Childhood, 58*(7), 535–537.

Bibring, G., Dwyer, T., Huntington, D., & Valenstein, A. (1961). A study of the psychological processes in pregnancy and of the earliest mother-child relationships: Some propositions and comments. In A. Freud, M. Kris, & E. Hartmann (Eds.), *Psychoanalytic study of the child* (Vol. 16, pp. 9–24). New York: International Universities Press.

Brust, J. E., Ford, C. V., & Rimoin, D. L. (1976). Psychiatric aspects of dwarfism. *American Journal of Psychiatry, 133,* 160–164.

Burlingham, D. (1972). *Psychoanalytic studies of the sighted and the blind.* New York: International Universities Press. •

Cohen, D., Dibble, E., & Grawe, J. (1977). Fathers' and mothers' perceptions of children's personality. *Archives of General Psychiatry, 34,* 480–487.

Drash, P. W. (1969). Psychologic counseling in dwarfism. In L. I. Gardner (Ed.), *Endocrine and genetic diseases of childhood* (pp. 1014–1022). Philadelphia: Saunders.

Emde, R. N., & Sorce, J. (1983). The rewards of infancy: Emotional availability and maternal referencing. In J. Call, E. Galenson, & R. Tyson (Eds.), *Frontiers in infant psychiatry* (pp. 17–30). New York: Basic Books.

Erikson, E. (1959). Identity and the life cycle. *Psychological Issues.* Monograph I. New York: International Universities Press.

Ferholt, J., Rotnem, D., Genel, M., Leonard M., Carey, M., & Hunter, D. (1985). Psychodynamic study of psychosomatic dwarfism: A syndrome of depression, personality disorder and impaired growth. *Journal of the American Academy of Child Psychiatry, 24*(2), 49–57.

Finley, B. S., Crouthamel, C. S., & Richmond, R. A. (1981). A psychosocial intervention program for children with short stature and their families. *Social Work in Health Care, 7*(1), 27–34.

Greenspan, S. (1981). *Clinical infant reports: Psychopathology and adaptation in infancy and childhood.* New York: International Universities Press.

Grew, R. S., Stabler, B., Williams, R. W., & Underwood, L. E. (1983). Facilitating patient understanding in the treatment of growth delay. *Clinical Pediatrics, 22,* 685–690.

Holmes, C. S., Hayford, J. T., & Thompson, R. G. (1982). Parents' and teachers' differing views of short children's behavior. *Child Care, Health and Development, 8*(6), 327–336.

Love, H. D. (1970). *Parental attitudes toward exceptional children.* Springfield, IL: Charles C Thomas.

McCollum, A. T. (1975). *Coping with prolonged health impairment in your child.* Boston: Little, Brown.

Meyer-Bahlburg, H. F. L. (1985). Psychosocial management of short stature. In D. Shaffer, A. A. Ehrhardt, & L. L. Greenhill (Eds.), *The clinical guide to child psychiatry* (pp. 110–135). New York: Free Press.

Parker, G. (1983). *Parental overprotection: A risk factor in psychosocial development.* New York: Grune & Stratton.

Parker, G., & Lipscombe, P. (1981). Influences on maternal overprotection. *British Journal of Psychiatry, 138,* 303–311.

Pruett, K. P. (1985). Disorders of the parent-child relationship. In J. Cavenar (Ed.), *Psychiatry.* New York: Lippincott.

Rotnem, D., Genel, M., Hintz, R. L., & Cohen, D. J. (1977). Personality development in children with growth hormone deficiency. *Journal of the American Academy of Child Psychiatry, 16,* 412–426.

Rotnem, D., Cohen, D., Hintz, R. L., & Genel, M. (1979). When treatment fails: Psychological sequelae of relative "treatment failure" with human growth hormone replacement. *Journal of the American Academy of Child Psychiatry, 19*(3), 505–520.

Sander, L. (1981). Investigation of the infant and its caregiving environment as a biological system. In S. Greenspan, G. Pollock (Eds.), *The course of life: Psychoanalytic contributions to understanding personality development: Vol. I. Infancy and early childhood* (p. 177). Washington, DC: Government Printing Office, OHHS Pub. No. (ADM).

Solnit, A. J., & Stark, M. H. (1961). Mourning and the birth of a defective child. In A. Freud, M. Kris, & E. Hartmann (Eds.), *Psychoanalytic Study of the Child* (Vol. 16, pp. 523–537). New York: International Universities Press.

Stabler, B., Whitt, J. K., Moreault, D. M., D'Ercole, A. J., & Underwood, L. E. (1980). Social judgements by children of short stature. *Psychological Reports, 46,* 743–746.

Weinberg, M. S. (1968). The problems of midgets and dwarfs and organizational remedies: A study of the Little People of America. *Journal of Health and Social Behavior. 9*(1), 65–71.

Winnicott, D. W. (1965). *The maturation process and the facilitating environment.* New York: International Universities Press.

Author Index

Numbers in italic indicate the page on which the full reference appears.

Subject Index